P. Gerhards, U. Bons, J. Sawazki,
J. Szigan, A. Wertmann

GC/MS in Clinical Chemistry

 WILEY-VCH

Further Reading from WILEY and WILEY-VCH

B. Kolb, L. S. Ettre
Static Headspace Gas Chromatography
1998. XX, 298 pp., Hardcover.
ISBN 0-471-19238-4.

K. Pfleger, H. Maurer, A. Weber
Mass Spectral and GC Data of Drugs, Poisons, Pesticides, Pollutants and Their Metabolites
Parts I - III
1992. 2nd ed. XXX, 3378 pp., Hardcover.
ISBN 3-527-26989-4.

G. Schomburg
Gas Chromatography. A Practical Course
1990. XIV, 318 pp., 126 figures and 8 tables. Hardcover
ISBN 3-527-27879-6

Petra Gerhards, Ulrich Bons,
Jürgen Sawazki, Jörg Szigan,
Albert Wertmann

GC/MS in
Clinical Chemistry

Translated by G. Leach

WILEY-VCH

Weinheim · New York · Chichester
Brisbane · Singapore · Toronto

Petra Gerhards
Shimadzu Europa GmbH
Albert-Hahn-Straße 6-10
D-47269 Duisburg

Ulrich Bons
Jürgen Sawazki
Apotheke der Rheinischen
Landesklinik Viersen
Johannisstraße 70
D-41749 Viersen

Jörg Szigan
Albert Wertmann
Labor Dr. Lembke und
Dr. Lempfrid
An der Wachsfabrik 25
D-50996 Köln

Library of Congress Card No. applied for.

A catalogue record for this book is available from the British Library.

Deutsche Bibliothek Cataloguing-in-Publication Data:
GC/MS in clinical chemistry / Petra Gerhards ... - Weinheim ;
New York ; Chichester ; Brisbane ; Singapore ; Toronto : Wiley-VCH, 1999
  ISBN 3-527-29623-9

Printing: betz-druck gmbh, D-64291 Darmstadt.
Bookbinding: J. Schäffer GmbH & Co. KG., D-67269 Grünstadt.
Printed in the Federal Republic of Germany.

Foreword

Developments over recent years indicate that isolated drug dependency is now the exception, while multiple abuse and addiction have become the rule. The drug abuser experiments with many substances with many and various effects, consuming them sometimes simultaneously and sometimes consecutively, sometimes even with deliberate "pharmacological matching". Powerful analytical techniques that are capable of detecting the broad spectrum of addictive substances must therefore be used. The sensitivity and specificity of some screening methods are often inadequate for the reliable determination of substances if these are consumed at low dose rates or are seldom used. An aspect which is becoming increasingly important is the analytical detection of co-consumption of all substances, a practice which endangers both the patient undergoing substitution therapy and the treatment itself.

The GC/MS method of analysis gives a large amount of information, especially in cases of multiple drug use. A good basic knowledge of GC/MS analysis, its power in clinical chemistry, the sample preparation process, the chemistry and pharmacokinetics of addictive substances, the possibilities for consumers or patients to tamper with samples, and the advantages and disadvantages of other analytical procedures are essential prerequisites for a competent interpretation of the results. Moreover, analysts engaged in drug screening must continually extend their knowledge to include new substances and/or abuse patterns.

Only with this broadly based knowledge can analysts do justice to their responsibilities, which include the social, health-related, and juristic consequences for the persons concerned.

Tübingen, November 1996   K.-A. Kovar

Preface

This book is written mainly for practical analysts. During many lecture tours we have been inspired by numerous conversations to set down the many and various possible applications of CG/MS in clinical chemistry in the form of a book.

The aim is to enable the reader to become proficient in GC/MS analysis as speedily as possible by explaining the fundamentals of gas chromatography and mass spectrometry, and to include the potential applications in the field of working medicine and drug analysis in a simple and practical form. Additional information from other areas of specialization is intended to help the analyst in the clinical laboratory to understand how these areas relate to his or her own work and to assess the analytical results.

Hence, when discussing "drug screening", the book offers information not only about GC/MS analysis but also about other analytical methods. Appropriate methods of sample preparation and quality assurance in the analysis are also discussed, and information on epidemiology and pharmacology is provided, without which an efficient drug screening is impossible.

Drug screening occupies a major part of this book, as the importance of this work continues to increase and GC/MS analysis has been the "gold standard" for many years.

Analysis by GC/MS has always been wrongly regarded as extremely costly, and we consider this point in the chapter "Internal Cost Calculation". We show that the costs of analysis depend mainly on the loading on the machine and that large synergy effects and low analysis costs can be achieved as a result of the wide range of potential applications of GC/MS in clinical chemistry.

Mönchengladbach, November 1996

Petra Gerhards
Ulrich Bons
Jürgen Sawazki
Jörg Szigan
Albert Wertmann

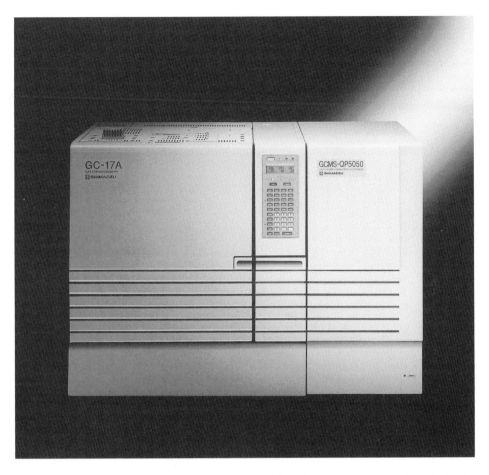

Modern Quadrupole GC/MS instrument used for the measurements in this book.
Shimadzu GC-17A and QP-5000/5050A series

Contents

Part I Fundamental Principles of Analysis by Gas Chromatography and Mass Spectrometry

Petra Gerhards, Jörg Szigan

1. Physical Theory and Equipment Design

Gas chromatography (GC) is based on the repeated partition or adsorption, between a mobile phase and a stationary phase, of components to be separated (Fig. 1-1). The mobile phase is always a gas, known as the carrier gas. The stationary phase can be either a solid or a liquid [1].

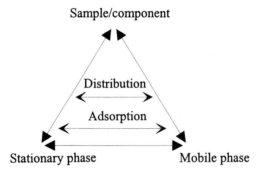

Fig. 1-1. Principle of GC separation

1.1 Adsorption

In adsorption chromatography, also known as GSC (gas-solid chromatography), the components are adsorbed on a solid, stationary phase. The stationary phase can consist of such materials as active carbon, silica gel, molecular sieves, aluminum oxide, or Porapack.

The principle of separation is based on the fact that the various components can be more or less strongly adsorbed by the adsorbent. In the adsorption process, the components of a gaseous or liquid sample are reversibly united with the surface of the adsorbent [1, 2].

1.2 Partition

In partition chromatography, also known as GLC (gas-liquid chromatography), the components are distributed between the two phases. Here, the stationary phase is a liquid, which may be directly applied to the walls of the column in the form of a thin film. Alternatively, the column may be packed with a solid material (silica gel, kieselguhr, Chromosorb, etc.) which is impregnated with the liquid.

For the separation of the individual components, the partition between the phases is important. On contact of the mobile phase with the stationary phase, a component i of one phase is transferred to the other (Fig. 1-2), and this leads to different concentrations of this component in the two phases. The ratio of the concentrations is given by the equilibrium constant K_i.

$$K_i = \frac{c_i(s)}{c_i(m)}$$

$c_i(s)$ – concentration of component i in the stationary phase
$c_i(m)$ – concentration of component i in the mobile phase

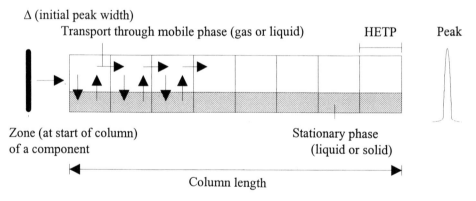

Fig. 1-2. Diffusion and transport of material in a chromatographic column

During transport through the mobile phase, the partition equilibrium is continually re-established in the various sections of the column along its length. One such short length of the column is known as a theoretical plate (HETP = height equivalent of a theoretical plate). The number of these plates is a measure of the separating power of a column. The number of theoretical plates will be large if the column sections of this type are short (small plate height) [1, 3].

1.3 Design of a Gas Chromatography System

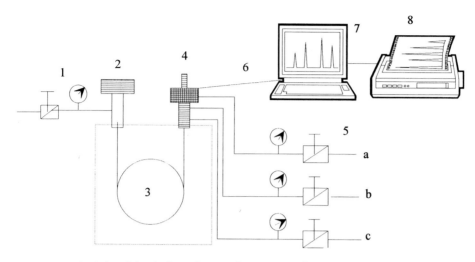

Fig. 1-3. Principle of the design of a gas chromatography system
     1 = carrier gas supply
     2 = injector (split or splitless)
     3 = column
     4 = detector (FID – flame ionization detector)
     5 = gas supply for the FID: a = air; b = fuel gas (hydrogen);
        c = purge gas (argon, helium or nitrogen)
     6 = serial port to computer
     7 = computer for display and integration of the chromatogram and control of
        the chromatography system
     8 = printer [3]

2. Injection and Headspace Technique

2.1 Injection

In gas chromatography, the samples must be introduced into the analysis system in the form of a vapor. Vaporization can be achieved either during or after introduction of the sample.

Liquid samples can be injected into the gas chromatograph by means of a microliter syringe, while solid samples must be dissolved prior to injection. The simplest method of vaporization is to use a hot injector system. Introduction of the sample is the only critical step in gas chromatography. Problems can include reaction of the components of the sample with each other or discrimination due to the components.

In principle, there are four types of sample injection:

- Splitless injection
- Split injection
- Temperature-programmed injection
- On-column sample injection.

2.1.1 Splitless Injection

This method is useful for very dilute solutions. When splitless injection is used, the column is overloaded with the solvent. For this reason, the temperature at the top of the column is kept low (10–20°C below the boiling point of the solvent), so that the low-volatility components and the solvent condense. This condensation causes the components to be focussed. The method is not recommended for volatile components, as these are eluted from the column with the solvent. Care must be taken in this procedure to prevent the injector from being overloaded by the quantity of liquid injected. The insert in the injector has an interior volume of ca. 0.4 ml. As the liquid is rapidly vaporized in the injector, the material can enter the cold upstream pipework if the amount of liquid is large, and condensation can occur. This can be a source of contamination in subsequent analyses [1, 3].

2.1.2 Split Injection

In the split injection technique (Fig. 2-1), only a part of the sample is delivered to the column. This method is used with capillary columns. Here, the sample is injected into the carrier gas stream through a septum, vaporized in the vaporizing tube, and then mixed with the carrier gas. The gas can be divided

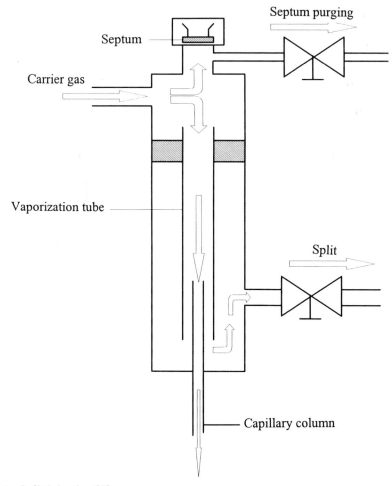

Fig. 2-1. Split injection [5]

into two streams by means of an infinitely adjustable needle valve, which should be adjusted so that a very small proportion of the sample is delivered to the column. This method is used when concentrations in the sample are high, as the capacity of a capillary column is low. During this process, the septum is continuously purged with carrier gas to prevent any substances that may be released from the septum itself (e.g. plasticizers) from reaching the column and interfering with the analysis. If the amount of sample applied is too large, the separating capacity of the capillary column is reduced because of overloading. The result is an inefficient separation. Capillary columns have a low flow rate, and a sample containing high concentrations of components must consequently be split to prevent a broad starting peak. On splitting, only a small proportion of the sample is delivered to the column, the rest being drawn off via the splitter valve. It is important to have a high flow rate through the injector, so that the sample reaches the column quickly. If the flow rate is low, the dwell time of the sample in the injector will be too great, and the starting peak will become too broad [3, 4].

2.1.3 Temperature-Programmed Injection

In this method of injection, the liquid sample is injected into a "cold" injector. After introduction of the sample, the injector is subjected to temperature-programmed heating, in which the split is programmed by an "EVENT-programme" (an auxiliary programme). Depending on the column temperature, a focussing effect at the top of the column can be produced. The advantage of this procedure is that the solvent can be removed by a time-controlled split before the actual separation takes place [1, 3].

This method is especially suitable for thermolabile components and for components which may form artifacts at high injector temperatures. An example of this is oxazepam. Use of a split/splitless injector leads to artifact formation, but this can be largely prevented by use of a temperature-programmable injector.

2.1.4 On-Column Injection

In this method, the sample is injected directly onto the column. Here, the sample is not contained in the glass insert. Columns with a small inside diameter are unsuitable for this technique of sample itroduction. As on-column injection is a splitless method, only low-concentration samples can be injected. This method is suitable for polar and thermally unstable components.

2.2 Headspace Technique

Gas chromatography (GC) in combination with the headspace technique (HS), known as headspace gas chromatography (HSGC), plays a dominant role in the investigation of volatile components in samples from clinical chemistry, biochemistry, food chemistry, and in environmental analysis. In contrast to the usual liquid injection, which is a direct analysis method, this is an indirect analysis method for the determination of volatile substances in liquid or solid samples, and is used when the usual direct methods for determining volatile substances in high-boiling, unstable, or low-volatility samples, especially in trace amounts, do not give satisfactory results.

This special injection technique has advantages over all other methods of analyzing volatile substances with respect to separating power, sensitivity, ease of handling, and automation possibilities, and the headspace technique is therefore becoming more and more widely used.

2.2.1 Static Headspace

In a static headspace, the sample is in a closed static system in which conditions are at thermodynamic equilibrium. After establishment of the equilibrium, the

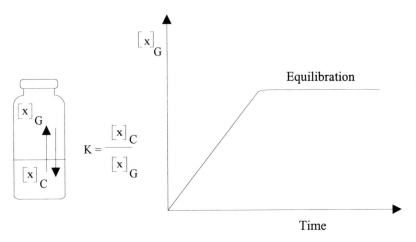

Fig. 2-2. Static headspace. $[x]_G$ = concentration of component x in the gas phase; $[x]_C$ = concentration of component x in the liquid phase; K = equilibrium constant

sample is removed directly from the vapor space above the sample and transferred to the gas chromatograph (Fig. 2-2).

For reproducible analysis, it is important that equilibrium between the phases should first be established. This equilibrium is influenced by the conditioning time and temperature. The removal of the sample for the GC analysis can be by means of overpressure or underpressure, although care must be taken that the temperature not only of the sample vessel but also of the metering device is kept constant. The HS vessels (vials) are closed with a PTFE (polytetrafluoroethylene) or aluminum-coated silicone septum. The septum itself is protected from the overpressure of the interior space by an aluminum cap. Before each analysis, a blank determination with an empty HS vial is carried out to detect whether any constituents of the materials of the septum or vial are being emitted.

2.2.2 Dynamic Headspace

In a dynamic headspace, there is continuous gas extraction of the volatile substance being analyzed. In this case, the concentration of the eluted substance changes continuously with time until it approaches zero asymptotically (Fig. 2-3). The dynamic headspace is thus independent of the matrix, as the whole amount is removed from the sample. The sample, liquid or solid, is in a vessel which is continuously purged with an inert gas (carrier gas). This method gives a greatly diluted gas stream, so that a concentration stage must be included before the actual analysis. This can be achieved by the use of an adsorption medium, a cooling stage, or a combination of both [6, 7].

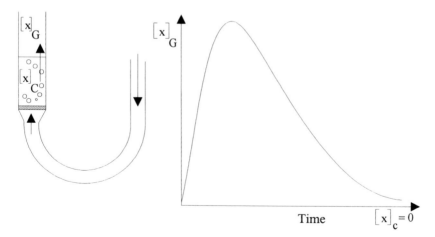

Fig. 2-3. Dynamic headspace. $[x]_G$ = concentration of component x in the gas phase; $[x]_C$ = concentration of component x in the liquid phase

3. Columns and Carrier Gas

Separation columns for partition chromatography are divided into two categories:

- Packed columns
- Capillary columns.

3.1 Packed Columns

These consist of glass or steel tubes with an inside diameter (ID) of 1–50 mm. Columns with an ID greater than 5 mm are used for preparative GC.

The column packings consist of an adsorbent solid (e.g. silica gel) with a surface area as large as possible and a fairly uniform grain size distribution. For partition chromatography, the packing acts as a carrier material and is impregnated with a liquid. Because of their large surface area, column packings can take up large amounts of liquid and are therefore particularly suitable for the separation of volatile components and gases [3].

3.2 Capillary Columns

These consist of fused silica (silicon dioxide), alkali glass or borate glass. Their ID varies between 30 and 500 µm. Column lengths are usually between 1 and 100 m.

There are three types of capillary column:

- *Packed capillary columns* in which the packing is an adsorbent (e.g. silica gel). These are used for adsorption chromatography only, not partition chromatography, and are suitable for strongly polar components.

- *Thin film capillary columns*, in which the liquid phase is applied to the inner walls of the column in the form of a thin film. Today, the liquid can also be chemically bonded to the wall of the column.
- *Thin coating capillary columns*, which consist of a thin, finely divided coating of carrier material which is then itself coated with the liquid phase [1, 2].

3.3 Stationary Phases

In partition chromatography, a wide range of compounds are used as stationary liquid phases:

- Polysiloxanes (silicones) with various nonpolar and polar substituents
- Polyethylene glycols
- Hydrocarbons
- Esters
- Polyesters.

The stationary phases can differ in polarity. Examples of the most commonly used phases are:

- SE-54 CB, nonpolar:

94% methylsilicone
5% phenylsilicone
1% vinylsilicone

Maximum temperature 300 – 320°C

- OV-1701 CB, medium polarity:

88% methylsilicone
7% phenylsilicone
5% cyanopropylsilicone

Maximum temperature 280°C

Carbowax, polar: $$\left[\text{O}-\text{CH}_2-\text{CH}_2-\text{O}\right]_n$$

Material with n > 500 is known as Carbowax 20 M.
When partially esterified with terephthalic acid or nitroterephthalic acid, it is known as WG 11 or FFAP (free fatty acid phase).

Maximum temperature 220°C.

All the above grades can be used in clinical chemistry. For drug analysis a thin film of an SE-54 grade, for example, is generally suitable, but for amphetamines in particular the use of a column of the OV-1701 type is recommended. Blood alcohol determinations can be carried out on a Carbowax column on account of the polarity of the components. The column type must be appropriate for the separation problems. All stationary liquid phases must fulfill certain requirements:

- They must have a low vapor pressure so that they do not pass into the mobile phase and affect the analytical results.
- There must be no chemical reaction with the components to be separated. This could prevent them from being eluted from the column or cause them to be chemically changed. Also, the separating power of the column is reduced.
- The stationary phases must be thermally stable over a wide temperature range, so that any breakdown products will not lead to false analytical results [6, 8].

3.4 Film Thickness

The uniformity of application and the thickness of the coating of stationary phase have a great influence on the quality and separating power of the column:

- Thin films: 0.10 – 0.15 µm for low-volatility samples
- Medium film thicknesses: 0.2 – 0.3 µm for standard separations
                                       0.5 mm for special separation problems
                                       As PCBs (polychlorinated biphenyls) 28/31
- Thicker films: 1 – 5 µm for volatile substances (blood alcohol)
                                       Large sample amounts (high loading capacity)
                                       Trace analysis (large concentration differences).

Table 3-1 gives data illustrating the influence of the film thickness and the inside diameter of the column on the capacity and phase ratio.

Table 3-1. Column parameters

Inside diameter (ID) [mm]	Film thickness [µm]	Phase ratio β	Capacity [ng/component]
0.32	0.1	800	60 – 70
0.32	0.25	320	150 – 200
0.32	0.5	160	250 – 300
0.32	1	80	400 – 450
0.32	5	16	2000 – 2500
0.53	1	133	1000 – 1200
0.53	5	27	5000 – 6000

The phase ratio β is calculated from

$$\beta = \frac{V_G}{V_L} \approx \frac{r}{2 \times df}$$

where r = radius of capillary column in µm
df = film thickness in µm.

The lower the phase ratio (β) between the volume of the mobile phase and that of the stationary phase the longer is the dwell time of the components in the stationary phase, i.e., the retention time increases. Longer dwell times can affect the peak shape. Peaks broadened by diffusion and adsorption lead to quantification problems [8].

3.5 Carrier Gas

In gas chromatography, the mobile phase is the gas. Gases used include nitrogen, helium, and hydrogen. The gases are supplied from a gas bottle and are fed via a pressure-reducing valve into the gas chromatograph. As they have different viscosities, the gases have different flow rates, and this affects the number of theoretical plates. Figure 3-1 illustrates that every gas has a minimum HETP at a certain flow rate. The number of theoretical plates is thus a maximum at this point.

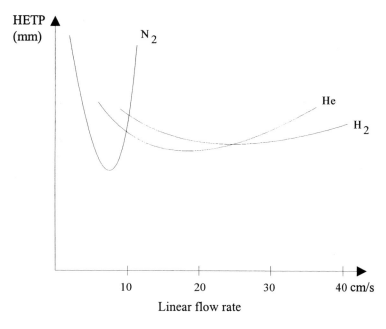

Fig. 3-1. HETP diagram

4. GC Detectors and Mass Spectrometry

The components eluted from the column are detected by the detector, which emits a proportional electrical signal, e.g., a voltage.

The detector of a gas chromatograph (Table 4-1) must have high sensitivity and reliability, simplicity of design, and a linear signal-concentration relationship over a wide range.

Table 4-1. Overview of GC detectors [8, 9]

Detector	Application/ selectivity	Limit of detection [g/s]	Linearity	Comments
Flame ionization detector (FID)	Almost all organic substances (C–C, C–H)	10^{-11}	10^7	Universal detector, simple operation
Thermal conductivity detector (TCD)	All	10^{-8}	10^4	Mainly for inorganic compounds
Photoionization detector (PID)	Halogenated aromatics and compounds with π-bonds	10^{-12}	10^7	Frequent problems with impurities
MS spectrometer IR spectrometer	Universal Universal			Also provide information on the structure of compounds
Electron capture detector (ECD)	Halogenated compounds and compounds containing nitro-groups	10^{-14}	10^4	For the detection of pesticides, polychlorinated biphenyls, and solvents. Extremely sensitive. Specified in official regulations

Table 4-1. (Contd.)

Detector	Application/ selectivity	Limit of detection [g/s]	Linearity	Comments
Flame thermoionization detector (FTD). Also thermoionization detector (TID) or phosphorus-nitrogen detector (PND)	Compounds containing phosphorus and nitrogen	10^{-13}	10^5	Important in food analysis. Detection of insecticides. Very selective, but rather complex in operation
Flame photometric detector (FPD)	Compounds containing phosphorus, tin or sulfur	10^{-11}	10^4	Each element requires its own special optical filter

4.1 Selectivity

This is a measure of the response characteristics towards the various compounds. Some detectors respond to almost all compounds and are referred to as "universal". Others only respond to certain types of compounds and enable the user to determine these compounds in a complex matrix. For this reason, GC detectors are divided into three groups:

- *Universal detectors* respond to all components. The thermal conductivity detector (TCD) is an example of this, as most eluates cause a change in thermal conductivity.
- *Selective detectors* can be selective towards elements, structure, or other properties. The flame ionization detector (FID) reacts selectively to substances that are ionized in a hydrogen/air flame (very broad selectivity). It is useful for the analysis of aqueous samples, as the water is not recorded. The electron capture detector (ECD) is selective for halogenated compounds.
- *Specific detectors* can detect particular structures or elements with a high degree of selectivity. Mass spectrometry that detects one ion only (SIM – selected ion monitoring) has high specificity. In SIM, individual fragments characteristic of the component being measured are detected.

4.2 The Flame Ionization Detector (FID)

Figure 4-1 shows the construction of a flame ionization detector.

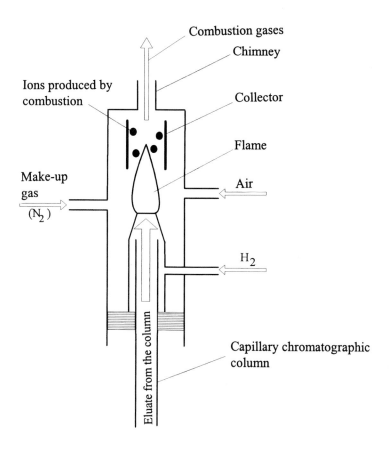

Fig. 4-1. Schematic design of a flame ionization detector (FID)

The FID responds to compounds that yield electrically charged species on combustion in a hydrogen/air flame, the free-radical reaction being

$$CH + O \rightarrow CHO^+ + e^-$$

These charged species, under the influence of an electric field, are captured on a collecting electrode and measured by an electrometer, whose output is amplified. The field of application of the FID is very large, as it responds to almost all organic compounds. A disadvantage is that it is often too unspecific and insensitive for environmental analysis and the analysis of residues.

4.3 The Electron Capture Detector (ECD)

The electron capture detector is especially suitable for halogen, sulfur, and nitro-compounds, i.e., compounds that are able to "capture" electrons. Because of its high selectivity and sensitivity, it is much used for residue analysis, e.g. for volatile halogenated hydrocarbons and plant protection agents. Figure 4-2 shows the design of an ECD.

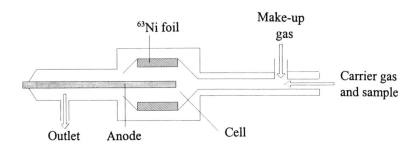

Fig. 4-2. Schematic diagram of an electron capture detector (ECD)

The beta-rays emitted from the cathode ionize the carrier gas, thereby liberating electrons. If a pulsed voltage is applied to the electrode in the cell, these electrons are captured, so producing an electric current. If electrophilic molecules are introduced into the cell, these absorb electrons and become negatively ionized. The electron density in the detector therefore decreases, so that a smaller number of electrons are captured at each pulse. The total number of electrons captured per unit of time (i.e. the current) can be kept constant by increasing the pulse frequency when the number of electrons decreases. The pulse frequency is then proportional to the concentration of the electrophilic molecules passing through the detector [8].

4.4 Mass Spectrometry

As a component is eluted from the column, it is introduced into the quadrupole mass spectrometer via an interface (direct or indirect). The component is ionized in the ion source of the mass spectrometer, and the ions formed are accelerated in the direction of the quadrupole to the detector (Fig. 4-3). The resulting distribution of molecular fragments is characteristic of the molecule.

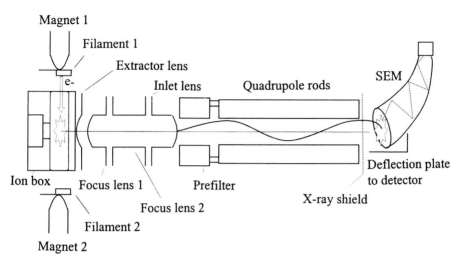

Fig. 4-3. Principle of a mass spectrometer

4.4.1 Design and Function of a Quadrupole Mass Spectrometer

A quadrupole consists of four parallel metal rods in a square configuration, two of which are connected together electrically. Ion separation takes place by deflection by an electrical field. An alternating voltage is applied to two oppositely located rods, producing alternating positive and negative fields relative to the central axis. Positive ions flowing through the rod system are accelerated toward the central axis during the positive phase and toward the rods during the negative phase. The extent of this lateral deflection from a straight line path depends on the applied voltage, the frequency (duration of influence of the alternating fields), and the mass of the ions. A positive constant voltage is applied in addition to the alternating voltage, and this causes a general deflection toward the central axis. In the case of heavy ions, the effect of the constant voltage predominates. Heavy ions are able to pass through the rod system, while light ions are deflected such that they strike the rods and are discharged. An alternating voltage whose phase angle differs by π from that of the other alternating voltage and also a negative constant voltage are applied to the other pair of rods. The negative voltage causes the ions to be deflected towards the rods to a certain extent, while, for low masses, the positive field of the alternating voltage is sufficient to bring them to the axis of the rod system. Thus, one pair of rods blocks off low masses and the other high masses. Suitable adjustment of the constant and alternating voltages ensures that only ions of a particular mass are able to pass through the rod system [10].

4.4.2 Detection

When ions strike the detector, they generate an electrical signal which is proportional to the number of ions. The secondary electron multiplier (SEM) is essentially a current amplifier. The ions strike the surface of a sheet of a special alloy, a "dynode", and cause electrons to be emitted from this. Because of the cascade design of the SEM, these electrons strike the next dynode and in turn cause more electrons to be emitted. A high voltage is applied to the first dynode, and, with the aid of a chain of resistors, voltages decreasing in magnitude are applied to all the subsequent dynodes. This gives a signal amplification of up to 10^6-fold. The output from the SEM is fed to an electrometer-amplifier, which converts the current into a voltage proportional to the measured value. This can be further electronically processed, and is normally digitized and stored by the data processing station.

4.4.3 Sample Introduction – the GC/MS Interface

When several components are ionized simultaneously, this leads to overlapping spectra. In order to obtain spectra of single substances, complex mixtures must be separated before they can be analyzed in the MS. The combination of GC and MS has the advantage that the components are already in vapor form and enter the MS separately. A capillary column with an inside diameter of 0.32 mm or less can be directly connected to the MS via the interface, which prevents the pressure due to the carrier gas in the ionization chamber from becoming too high. Megabore capillaries (I.D. 0.53 mm) and packed columns require a special interface.

A direct sample introduction system is suitable for pure liquids and solids. Here, the sample is placed on the point of a movable rod which can be introduced into the ion source through a special valve. The sample is then vaporized and broken down into ionized fragments.

4.4.4 The Total Ion Current (TIC)

The total ion current (TIC) is the current due to the total number of ions passing through the analyzer (Fig. 4-4). It is necessary to measure this because part of the total ion stream is removed and discharged on the quadrupole rods. The remainder, i.e. the part of interest, reaches the detector, where it is amplified and converted into an electrical signal, which is fed to the data processing station. The TIC is a measure of the ion yield, and hence indirectly also of the vapor pressure in the source.

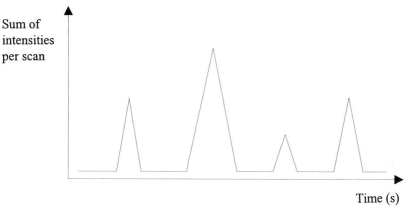

Fig. 4-4. Total ion current (TIC) [8]

To identify the compound, the signal is split up to show the fractions of the individual masses, i.e. a mass spectrum (Fig. 4-5).

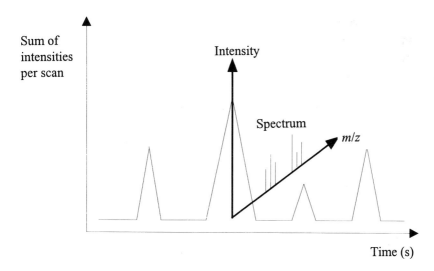

Fig. 4-5. Splitting in a mass spectrum. m/z = mass/charge ratio [8]

4.4.5 Selected Ion Monitoring (SIM)

In drug analysis, it is often necessary to analyze complex mixtures. However, often only certain components are of interest. For these components, selected masses (molecule-ions and characteristic fragments) are recorded during the complete GC analysis as one mass after another is selected in a cyclic manner. Only in the case of the GC peak that contains the component of interest (the target compound) is the signal of all the selected ions in the correct intensity ratio observed. The gas-chromatographic retention time can also be used as an additional criterion of identity.

4.4.6 Data Processing

In most cases, all ions are recorded with their relative intensities. In the presentation of the data, the ion with the highest intensity (the "base peak") is usually assigned a value of 100, and the other ion intensities are given as a percentage of this (% relative intensity).

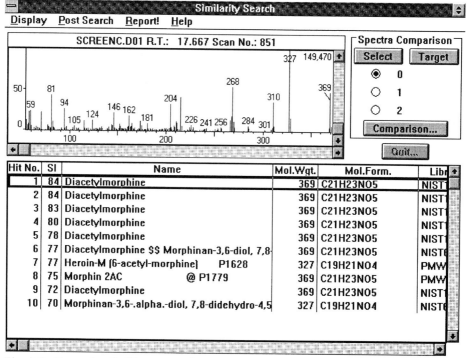

Fig. 4-6. Example of a library search

The interpretation of spectra is considerably facilitated by modern data processing. In commercial libraries, e.g. the NIST 75.000 or the Pfleger-Maurer-Weber library for drugs and pesticides, reference spectra are available that can be compared with the measured spectra, the degree to which the measured spectrum matches the entry in the library being known as the "similarity". The library search can be fully automated (Fig. 4-6), which represents a considerable saving in time.

Today it is also possible to use private libraries. These can be extremely useful, especially in the area of drug analysis.

A library also facilitates the identification of the individual components. A suitable preliminary search can enable the characteristic masses to be determined and placed under the TIC chromatogram as mass fragments (Fig. 4-7).

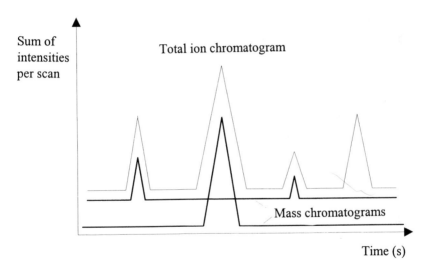

Fig. 4-7. Mass fragments under a total ion chromatogram [8]

Commercial libraries contain additional information, e.g., structures, molecular masses, and, if applicable, trade names of compounds (Fig. 4-8).

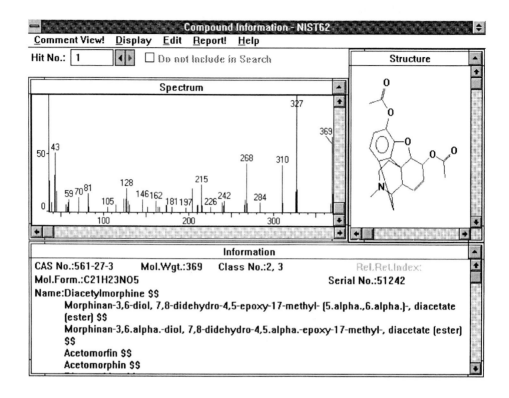

Fig. 4-8. Information typically available in a commercial library

5. Quantification

5.1 Methods Based on Internal Standards

In evaluation using an internal standard, a known amount of another component is mixed with the sample. This component must fulfill the following conditions:

- It must not be already present in the sample.
- It must be chemically similar to the target compound.
- It must have similar retention properties.
- It must not react with or chemically change the sample.

Quantification is performed by comparing the standard peak area with the sample peak area. A factor (the response factor) between the sample and the standard must be derived, as equal amounts of sample and standard are detected by the detector with very different sensitivities.

The response factor of component i is calculated from

$$ f_i = \frac{A_{st} \times G_i}{A_i \times G_{st}} $$

f_i — response factor of the component
G_i — weighed amount of the component
G_{st} — weighed amount of the standard
A_{st} — peak area of the standard
A_i — peak area of the component.

The component i itself is calculated from

$$ \text{Component } i = \frac{A_i \times f_i \times Z_{st}}{E \times f_{st} \times A_{st}} $$

E – weighed amount of the sample without internal standard
f_{st} – response factor of the internal standard
Z_{st} – weighed amount of the internal standard.

The disadvantage of this method is that an internal standard is necessary for each component group.

5.2 Standard Addition

Here, a definite amount of the component to be quantified is added. The change in peak area caused by the increase in concentration is used to quantify the component. Improved quantification is obtained if various concentrations are added to the sample, so that a straight line is obtained when the peak areas are plotted against the concentrations (Fig. 5-1).

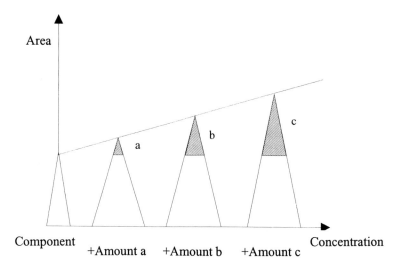

Fig. 5-1. Example of standard addition. a, b and c represent concentrations of a
      component added to the sample

A straight line can be obtained from the results (linear regression). The regression coefficient r^2 gives a measure of the linearity. The closer r^2 is to unity, the more accurately is the relationship between X and Y represented by a straight line.

The concentration of the sample by the standard addition method is calculated from:

$$c = \frac{c_{st} \times A_0}{A_{st} - A_0}$$

c – concentration of the sample
c_{st} – concentration of the standard
A_0 – peak area of the component in the sample without addition
A_{st} – peak area of the component in the sample with standard.

5.3 External Calibration

In this method, the components to be determined are added individually in definite amounts. The peak areas obtained are then plotted against the concentration. This method is also used for the preparation of a calibration curve. The disadvantages of the method are that production of the curves is time-consuming and matrix effects are ignored [1–3].

5.4 Quantification in the Split Mode

In analysis in the split mode, improvements can be obtained if certain errors are excluded from the outset. A split flow can be pre-set to give a predetermined gas flow rate at the inlet to the column, but the true split flow is often different from the "pre-set" value. In many cases, more sample material enters the column than the "pre-set" amount.

The rapid vaporization of the liquid sample in the vaporization chamber resembles an explosion. In a short time, a gas volume is produced in the injector that is often larger than the volume of the injector.

The increase in the pressure at the inlet of the column causes an increase in the flow rate in the column (Period A). When the pressure returns to its normal value, the flow rate in the column decreases. Sometimes the flow direction reverses momentarily if the pressure at the inlet of the column is higher than that in the injector. The flow rate through the split valve changes in an analogous manner but for a different time and to a different extent. This is important for the splitting of the sample when the "cloud" of sample vapor reaches the split point.

The injection of 2 µl liquid gives a cloud of vapor with a volume of ca. 0.5 ml. Consequently the pressure wave, starting from the center of vaporization below the outlet of the hypodermic needle, affects the gas flow rate into the column and through the split valve. The split ratio is obtained using this relationship. Figure 5-2 shows the change of pressure at the inlet to the column and the flow rate in the column. The split flow rate behaves similarly. The pressure wave is "sharp" when it passes through the column inlet and broad when it passes through the split valve.

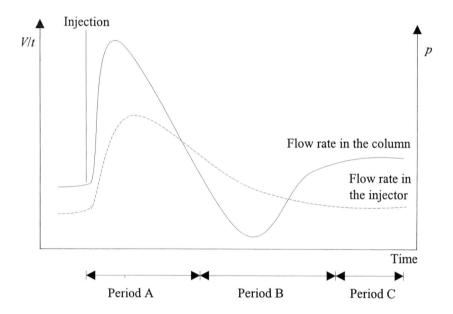

Fig. 5-2. Schematic diagram of the changes of pressure and flow rate during injection caused by the explosive expansion of the sample during vaporization [11]

The flow rate through the column and the split valve changes in various ways, so that the split ratio varies. The effect of the sample splitting depends on the moment at which the cloud of vapor passes the split point. If the needle discharges the sample near to the column inlet, the pressure wave delivers most of the sample vapor into the column, but if the sample is discharged at a point above the column inlet, the cloud of vapor passes the split point just when the pressure drop begins (Period B). The pressure at the inlet of the column is then relatively high, and only a small proportion of the vapor enters the column. Thus, the pressure wave generated when the sample vaporizes changes the carrier gas flow rate at the moment when the sample is split.

Many samples are focussed at the column inlet because of recondensation. This occurs at the point of transition from the heated injector to the column, which is 10-20°C cooler than the boiling point of the solvent. If no solvent is present, the components themselves recondense. If the sample vapor recondenses at the column inlet, this leads to a drastic reduction in volume. The pressure at the recondensation point falls because more sample material is drawn into the column.

To quantify the split flow, three criteria must be fulfilled:

- The peak areas must be proportional to the concentration in the sample.
- The relative sizes of the peaks must be identical with the previously calculated values or with relative areas found during splitless injection.
- The relative sizes of the peaks in the chromatogram must remain constant, even if the analytical conditions are changed (temperature, split ratio, flow rates, etc.).

The term "linearity" is used today to indicate that the split ratio is identical for all sample components. This is the basic precondition for ensuring that the small amount of sample material analyzed by the column has the same composition as the sample in the injector. However, it does not mean that the true split flow must be the same as the "pre-set" split flow. A complementary concept is the idea of discrimination, which is the opposite of linearity. Discrimination is not a very important effect in the headspace technique as the sample is already in vapor form, but it is considerably more significant in liquid injection [11].

5.5 Limits of Detection

The signal-to-noise ratio S/N is a measure of the lowest detectable concentration of a substance. Only when the difference ΔS between the measured value P and the average value of the noise N is three times as large as the standard deviation s of the noise can it be stated with 99% certainty that a measured value P in the chromatogram can be regarded as a signal and not as a component of the noise.

The limit of detection (Fig. 5-3) is used to describe the sensitivity of a detector [5].

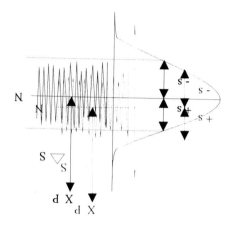

Fig. 5-3. Limits of detection: the condition $\Delta S = P - N \geq 3s$

References for Part I

[1] Schlegelmilch, F., *Vorlesungsscript Instrumentelle Analytik*, Fachhochschule Niederrhein, Krefeld, September 1990.

[2] Günther, W., Schlegelmilch, F., *Gaschromatographie mit Kapillar-Trennsäulen*, Band 1, Würzburg: Vogel Verlag, 1986.

[3] Schomburg, G., *Gaschromatographie*, 2. Aufl., Weinheim: VCH Verlag, 1986.

[4] Sandra, P., *Sample Introduction in Capillary Gas Chromatography*, Volume 1, Heidelberg: Huethig Verlag, 1985.

[5] Szigan, J., Diplomarbeit, Fachhochschule Niederrhein, Krefeld, 1993/94.

[6] Schlegelmilch, F., *Analytik flüchtiger Schadstoffe in Umweltproben*, FHN-Broschüre Fachhoschule, Niederrhein, Krefeld, 1990.

[7] Hachenberg, H., *Die Headspace-Gaschromatographie als Analysen- und Messmethode*, 2. Aufl., Hoechst, 1988.

[8] Geissler, M., *Einführung in die Gaschromatographie*, Shimadzu Europa GmbH, Duisburg, 1993.

[9] Naumann, H., Heller, W., *Untersuchungsmethoden in der Chemie*, 2. Aufl., Stuttgart: Thieme Verlag, 1990.

[10] Budzikiewicz, H., *Massenspektrometrie: Eine Einführung*, 3. erweiterte Aufl., Weinheim: VCH Verlag, 1992.

[11] Grubb, C., *Quantitation by Split Injection, Classical Split and Splitless Injection in Capillary GC*, Heidelberg: Huethig Verlag, 1988.

Part II Drug Screening

6. Epidemiology of the Abuse of Drugs and Medicaments

Ulrich Bons

6.1 Introduction

As long as people abuse drugs and medicaments, there will always be a need to analyze for these materials to aid diagnosis and point towards a possible therapy, e.g., in-patient detoxification. Also, in the field of substitution therapy with methadone, narcotics legislation and the guidelines of the German medical council (NUB) prescribe continuous control analyses, in particular to detect and exclude the consumption of other drugs or medicaments at the same time as the methadone, a practice that can cause reactions that can greatly endanger the health of the patient. Control analyses can also be necessary, for example, to comply with regulations of legal or medical organizations.

The analyst's awareness of which substances (some quite new) are being consumed and in what form, and which examples of abuse are in vogue, necessarily lags behind that of the consumers. Drug addicts are not lacking in ideas for circumventing supply problems by using other substances or by changing consumption habits in other ways. The analyst can therefore often only react after the drug scene has acted. It is self-evident that powerful analytical methods must be used to cope with the broad spectrum of drugs of abuse as fully as possible. If the sensitivity and specificity of the screening methods used are not capable of detecting even minute consumed doses of little-used materials and the addict is consequently not convinced of the effectiveness of these methods, news of this spreads rapidly among the patients concerned.

Analytical results, with their frequently personal, juristic, social and economic consequences for the person to whom they relate, are very important, and stringent demands are made not only on the apparatus in the investigating laboratory but also on the investigating personnel themselves. A knowledge of the chemistry and pharmacokinetics, including metabolism and excretion, of the addictive substances, therapeutic medication, methods of sample preparation, and the specificity and limits of detection of the chosen analytical procedure have a decisive influence on the quality of the analytical results. The method ultimately chosen in a particular instance depends on the epidemiology, previous history, and status of the patient.

6.2 Specific Data and Trends

Almost daily we read press reports of record-breaking quantities of drugs used or of the dramatic reduction in the age at which drug abuse starts. In these reports, certain substances are always at the forefront of popular awareness while many others are hardly ever mentioned. For example, press comments on the relevant annual reports for 1994 highlighted the consumption of amphetamine derivatives, but detailed information on the LSD market, which was developing in parallel, could in general only be obtained from the reports themselves.

None of the widely quoted statistics embrace all those individuals dependent on drugs and medicaments, but always only those who attract attention because of crime, their lifestyle, or their social environment. In the case of benzodiazepines, appetite-suppressants and combinations of analgesics with caffeine, the low-dose dependence of the patient is often associated with social conditions, so that the problem is often seen in the context of the particular decade in which the abuse occurs. The same is true of the experimental drug-taker who discontinues his or her consumption after a short time and is not noticed. For this reason, exact national statistics on consumption behavior are difficult to obtain. Moreover, it is possible to detect trends, e.g. the appearance of new dependence phenomena, patterns of abuse, drug combinations or methods of application, and the analyst must arrange his work accordingly.

6.2.1 Opiates/Opioids

6.2.1.1 Heroin

Heroin has a special place among the opiates, as most of our practical knowledge of drug dependence has come from experience with this drug. The behavior of heroin addicts attracts more attention than that of other drug users and is more likely to lead to crime and to committal for therapy. In most statistics, heroin users feature disproportionately in comparison to users of cannabis or benzodiazepines. Heroin users are in most cases polytoxicomanes, also using substances such as cannabis, codeine, cocaine, benzo-diazepines and/or alcohol. Other drugs such as sedative antidepressants (doxepin, amitriptyline, etc.) or antihistamines (diphenhydramine, doxylamine) are often also used. The starting age for heroin is, on average, ca. 20 years [1].

Heroin can be obtained on the illicit drugs market in the form of the base, and it is dissolved with ascorbic acid or citric acid with gentle heating before it is injected. The active substance content of heroin packets is not quoted, but varies between ca. 1% and 80%, an average figure being ca. 20% [2]. This ignorance of the undeclared dilution of the drug therefore often leads to overdosing, which is often fatal. "Cutting" of the drug is done with caffeine, paracetamol, codeine or diazepam. In the drug scene, a mixture of heroin and cocaine is known as "speedball" [3].

Heroin can sometimes be very pure and can then be smoked, a practice that is increasingly replacing injection. Many young people are meanwhile switching to other substances such as amphetamine and its derivatives or to cocaine, whose consumption is said to produce a more intense "high".

6.2.1.2 Codeine

The habit-forming potential of codeine in comparable antitussive doses is considerably lower than that of morphine, but the codeine dose can be increased so much in extreme cases that an effect comparable to a heroin "high" can be produced. Average daily dose rates are between 500 and 750 mg, but can be increased tenfold in an extreme case. Most codeine users show dependence behavior, taking the drug regularly over a long period not merely to get over heroin supply problems, although the drug is often taken in combination with heroin [4]. The starting age (which is tending to rise) for codeine abuse is somewhat higher than that for heroin, i.e. 23–24 years. If the medicinal preparation consisting of codeine combined with paracetamol or acetylsalicylic acid and containing a comparatively low dose of codeine phosphate hemihydrate (10–50 mg) is used to achieve the above-mentioned codeine dose rates of 750–7500 mg/day, there is a high risk of intoxication due to the 10–15 g/day of peripheral analgesics which are also taken.

6.2.1.3 Dihydrocodeine (DHC)

The epidemiological data for DHC is analogous to that for codeine, although this substance is more favored than codeine by drug users. In the frequently practiced DHC substitution therapy, a 2.5% solution is used, but patients in a few cases have increased the usual daily dose from the 300–500 mg range to as much as 5000 mg, adding alcohol as an "accelerator" and other psychotropic substances such as heroin, benzodiazepines and cocaine. Substitution without use of other substances is rare [5].

6.2.1.4 Methadone

A change in the narcotics regulations (BtmVV) on 01.02.93 first allowed the use of methadone in Germany as a centrally active analgesic for opiate substitution. The oral use of methadone can suppress withdrawal symptoms, but, in contrast to heroin, does not give any euphoric effect: there is no "high". However, this sensation can be induced by intravenous application and/or combination with alcohol, benzodiazepines, etc. Therefore, the legal authorities require that the substance should be supplied in a form that cannot be applied parenterally (addition of viscous and acid-containing substances) and that regular urine testing should be carried out.

   The illicit drugs market deals in methadone just like DHC, but abuse is less often observed than with the three opiates mentioned above [6]. As it is also available without the substances added to prevent parenteral administration, abuse by intravenous application is relatively common.

6.2.1.5 Tramadol

Apart from Valoron N®, this substance is the only centrally active analgesic that is not prohibited by the BtmVV. Abusers are sometimes found among medicament-dependent patients who additionally consume other substances such as peripherally active analgesics, benzodiazepines or antidepressants. This substance is not common in the drug scene, and frequency of abuse shows large regional variations.

6.2.1.6 Tilidine

The abuse of Valoron®, which was formerly widespread, decreased to an insignificant level after the single-substance preparation became subject to the BtmVV reulations (01.04.1978). The grade of Valoron® now available contains not only tilidine but also the opiate antagonist naloxone, making intravenous application almost impossible. Furthermore, it still to some extent retains its old "image" as a temporary aid in the drug scene, and the number of users is correspondingly low.

6.2.1.7 Other Centrally Active Analgesics

Fentanyl and its derivatives are currently seldom abused, or at least such abuse tends not to be noticed. The same applies to pethidine and the other opiates/opioids. However, changes in the drug scene could quickly lead to a change here also.

6.2.2 Peripheral Analgesics

The abuse of peripherally active analgesics and combinations of these, especially with caffeine, is well enough known, although concrete quantitative data are hard to find. The ready availability of these substances, which can be obtained for self-medication in every pharmacy, results in their widespread consumption. Approximately 70% of users prefer combinations with caffeine to single-substance preparations. Here, the combination of the caffeine with the euphoric effect of the analgesics is decisive for the abuse. Both groups of preparations are also used by alcohol-dependent people to treat morning hangovers.

Daily doses of 10–20 g (40 g) based on paracetamol or acetylsalicylic acid (aspirin), also in combination with alcohol, are often kept up over several years by alcohol- or medicament-dependent people. These substances are not of great importance for drug users. Sometimes phenacetin is consumed as this, like aspirin and paracetamol, has a euphoric effect in high doses, but this substance has not been sold in Germany since 1987. Metamizol, ibuprofen and propylphenazone are also relevant.

The habit of abusing peripherally active analgesics is often established at an early age. Later, appetite suppressants, amphetamines as stimulants, and benzodiazepines or laxatives may be taken. Peripheral analgesics therefore play a not inconsiderable role as starter drugs, and there is a special danger of dose increase due to addiction and chronic abuse.

6.2.3 Benzodiazepines

Benzodiazepines, introduced in the early 1960s, have almost completely displaced barbiturates and bromurides from the list of prescribed drugs. Because of their better compatibility, fewer side effects and lower toxicity, they represented an innovative contribution to the treatment of mainly psychic disorders [7].

The problems of benzodiazepine abuse are well enough known and are as old as this group of substances itself. All benzodiazepines have the potential to be addictive and represent some of the most commonly prescribed medicaments worldwide. Their abuse is therefore highly relevant epidemiologically. At a conservative estimate, ca. 1 000 000 people are addicted to benzodiazepines in Germany alone. This means that the abuse of these substances comes in third place after alcohol and nicotine. They are often the first substances of abuse for young people, coming *before* alcohol and nicotine [8]. Out of this appalling total, the main substances, which patients suffering low-dose dependence consume in therapeutic doses, sometimes over decades, are bromazepam, diazepam, dipotassium chlorazepate, flunitrazepam, lorazepam and oxazepam. Hangover effects due to the often slow elimination of substances and their metabolites prompt the taking of stimulants on the "morning after" [9]. Benzodiazepines abused by polytoxicomanes include diazepam, flunitrazepam and bromazepam.

The differences between the various diazepines with respect to the nature and duration of the effects of they produce are obviously of little relevance to the abuse of these substances.

6.2.3.1 Diazepam

Diazepam is the most abused substance of all the benzodiazepines. Not only drug addicts, who use the substance as a bridging aid and to intensify other drug effects, have experience of diazepam, but so also do most medicament-dependent people and alcoholics. The age of first use is lower than in the case of other benzodiazepines, sometimes as early as childhood. Diazepam is therefore also important in the introduction of drugs to people who later exhibit mono- or polytoxicomania.

The age of first use varies over a wide range, the average being 20–25 years, and, with the exception of a small number of experimental users, most patients abuse this substance over a period of at least 3 years. In a few cases, the duration of dependence exceeds 25 years [1].

6.2.3.2 Flunitrazepam

Flunitrazepan is very well known in the area of general abuse of benzodiazepines. It is mainly taken by polytoxicomanes, who also use heroin, DHC, cocaine etc., usually in doses of up to 20 mg/day. The extent of the abuse of this drug among this group remains at a high level, and no decrease is yet in sight. In the drug scene, flunitrazepan is marketed as a hard drug. The combination with heroin and cocaine is known as "speedball cocktail".

Limitation of the dose of this drug (since 01.04.94) to 1 mg per unit of medication (e.g. tablet) has hardly changed the abuse pattern. Dependents adjust the number of units of medication taken according to the "feeling", although the problems of procurement are multiplied accordingly.

6.2.3.3 Bromazepam

Abuse of the other benzodiazepines is typified by bromazepam abuse, which is in third place in the abuse statistics. It is consumed in similarly increasing amounts by all patient groups (drug, medicament, and alcohol abusers). Although a few individual dose rates reach 100 mg, most users do not attract attention or are detected too late because of low dose rates. Drug users also combine the substance with other benzodiazepines.

6.2.4 Barbiturates

Barbiturates now no longer play any part in the treatment of sleep disturbance and psychiatric problems, as in most cases they can now be replaced by better tolerated substances with a more favorable side-effect profile. Only phenobarbital is of therapeutic value in the treatment of epilepsy, although thiopental and methohexital are still used to produce general anesthesia. Almost all other barbiturates have been withdrawn from the market and are subject to the legislation covering narcotics formulations. For this reason, barbiturates ("lallas") are seldom abused, and only then by a small group who seek not only to experience their opiate effects but also to distance themselves from their surroundings. Because of this change in demand, there has been a shift in consumption in favor of sedative antidepressants such as amitriptyline and doxepin.

6.2.5 Other Sedatives and Addictive Substances

6.2.5.1 Chlormethiazole

The abuse of this substance often follows therapeutic use (treatment of alcoholism or delirium). Consequently, the patients come from the circle of alcohol abusers or people dependent on both alcohol and medicaments. The transition to abuse of this substance, also in combination with other medicaments, occurs at an average age of 33–34 years. The dose can range from therapeutic levels up to a maximum exceeding 25 g/day. The abuse of chlormethiazole shows no sign of decreasing. Dose rates and duration of abuse still remain at a very high level [1].

6.2.5.2 Antihistamines

Diphenhydramine is consumed as well as doxylamine to a small extent. Both substances are also found as "intensifiers" in plant-derived hypnotic drugs used in self-medication. Sedative antihistamines are regularly consumed by alcoholics and medicament-dependent people over long periods. The dependence of most patients does not tend to attract clinical attention.

6.2.6 Amphetamines and Related Substances

6.2.6.1 Substances of the Drug Scene

The abuse of amphetamine and its derivatives has recently undergone a virtual rebirth. Many unscrupulous chemists worldwide are taking part in the production of "speed", "rosenspeed" and "edelkoks" (the most commonly used names), which contain amphetamine or methamphetamine in widely varying concentrations according to the country of origin. Also, there is a continually increasing use of the so-called techno-drugs of the ecstasy group (see Fig. 6-1): 3,4-methylenedioxy-ethylamphetamine = MDE, 3,4-methylenedioxyamphet-amine = MDA and 3,4-methylenedioxymethamphetamine = MDMA ("ecstasy" = "XTC", "Adam", "Eve" etc.) [10]. These substances are supplied in tablets of various sizes, colors and pressed densities. The usual patterns embossed on the tablets (hammer and sickle, diamond shape, ADAM, Popeye etc.) act as a kind of trade mark, but do not allow any conclusions about the actual composition to be drawn.

In parallel with the growing quantities of amphetamines used, the abuse of these newer products is also increasing. Principally young people, especially participants in the techno-scene, use their hallucinogenic and stimulant effects [11]. The duration of abuse seldom exceeds 1 year. Patients mainly do not attract attention, as consumption takes place sporadically and for limited periods. The proportion of experimental users who turn away from the drug after a short period is high. Those drug users who in the past consumed only benzodiazepines as proprietary medicaments along with the illegal drugs derived from plants (heroin, cannabis and cocaine) are now also using substances of this group to an increasing extent, and the further spreading of this drug abuse has not yet ceased.

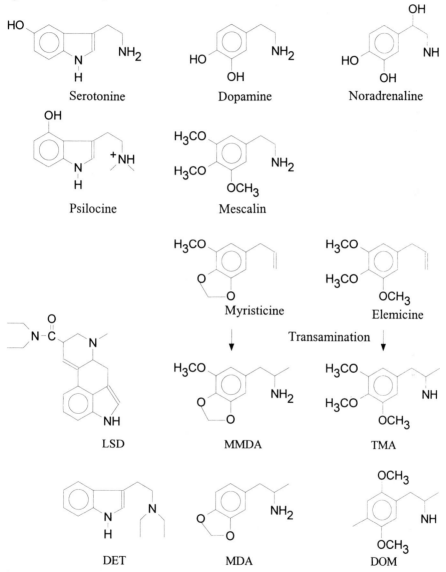

Fig. 6-1. Structural relationships between neurotransmitters and some plant-based and synthetic hallucinogens

"Look-alikes" are also on offer as tablets or powder which contain no narcotic agent but whose appearance resembles that of illegal products. Substances sold as substitutes for cocaine and amphetamines include, e.g., mixtures containing high doses of caffeine.

6.2.6.2 Psychoanaleptics and Antihypotonics

The amphetamines with a primarily psychoanaleptic effect are mainly subject to the narcotics laws or are no longer on the market. Medicaments such as fenetyllin (the substance most abused by young people in the mid-1980s), amfetaminil and methylphenidate are no longer used in significant quantities. Prolintane, marketed as an antihypotonic until 1994, is still occasionally used. However, the likelihood of encountering any of these today is fairly remote.

6.2.6.3 Appetite Suppressants

The abuse of appetite suppressants is widespread, but concrete numerical data are difficult to obtain as most abusers do not attract attention clinically. The eating disorders leading to this abuse also show themselves in the consumption of laxatives and diuretics. Some patients combine appetite suppressants such as norephedrine, norpseudoephedrine or amfepramon with benzodiazepines. The sympathomimetic effect of these substances can compensate in these cases for the morning hangover effects of the tranquilizers.

Harmful slimming formulations, which still appear from time to time year after year, often contain, as well as amfepramon and fenfluramine, benzodiazepines and diuretics. The current and usually not apparent abuse of these products was brought to the attention of the public in the summer of 1995 when the media drew attention to the use by patients of these dangerous products for weight reduction over long periods, especially in the Cologne-Aachen region. The period of use of permitted proprietary medication also consistently considerably exceeds the medically recommended 4-week period for supplementary therapy for the treatment of diet-related excess weight.

6.2.7 Designer Drugs

The term designer drugs is used mainly by the popular press to describe, often falsely, all synthetic drugs, especially the second generation amphetamines. Correctly defined, it describes analogs of substances that are controlled by being subject to legal restrictions [12]. Chemical manipulation (methylation, chain

extension etc.) changes the parent substances so that a new chemical substance is formed which is no longer subject to the narcotics laws but has undergone only small changes in its pharmacodynamic effects.
The principal substances of this type are:

- Amphetamine derivatives excluding "ecstasy"
- Tryptamine derivatives
- Phencyclidine derivatives
- Fentanyl derivatives (e.g. α-methylfentanyl = "china white")
- Pethidine derivatives = prodine

Quantitatively, the abuse of these substances is not very significant in Germany. However, this situation could change at any time.

6.2.8 Cannabis

In the Federal Republic of Germany, cannabis is mainly consumed in the form of the dried resin ("hashish"). Cannabis is the most important drug from the point of view of duration and extent of abuse. Almost all drug abusers consume this substance, mainly by smoking [13], along with opiates and tranquilizers. However, many young people who are only mild drug abuseres have experience of cannabis. The proportion of regular consumers is very high, and consequently the proportion of occasional experimental users is very low. Although cannabis is relevant as an introductory drug, its consumption does not necessarily to lead to a career of drug abuse, although the way is undoubtedly prepared. A number of other substances can also act as an introduction to general drug taking. Thus, cannabis is in third place after alcohol and nicotine in the list of drugs of introduction.

The average age of introduction to cannabis consumption is 15–17 years, and the duration of abuse is, on average, 6–7 years [1].

6.2.9 Cocaine

In the past, cocaine was always something special because of its high price. Also, the wave of crack-taking in the 1980s hardly touched the Federal Republic of Germany. However, this situation has changed greatly as a result of the falling world price of cocaine [14]. The South American cartels have taken over the European market, and sales are rising.

Most cocaine consumers are polytoxicomanes. One seldom encounters people who inhale or smoke cocaine only. Cocaine is generally taken by sniffing, but intravenous injection does occur. Introduction to cocaine consumption takes place at an age between 15 and 30 years, and the proportion of experimental users is rather high at ca. 20%, as with all stimulants. The proportion of regular cocaine consumers is certainly decreasing to some extent, although cocaine competes with other stimulants whose total consumption has increased. The number of first consumers of cocaine in Germany continues to rise, the increase in 1994 being 33.2%.

As amphetamine derivatives and cocaine are often used as alternatives, any increase in the supply of these substances favors increased cocaine abuse. Procurement problems decrease when amounts of imported material increase. So far, synthetic substances produced in Germany are cheaper.

6.2.10 Other Hallucinogens

The group of hallucinogens includes 40–50 substances, including their different forms. These are very heterogeneous in their chemical structure. Not only substances with the basic indole structure such as LSD and the secale alkaloids, but also parasympatholytic substances (piperidine derivatives), phenylethyl-amines such as mescalin, substances present in nutmeg, isoxazole derivatives (muscimol) and quinoline alkaloids (aconitine) have hallucinogenic effects. We must also include the synthetic substances of the amphetamine type, the designer drugs and some medicaments used in the treatment of Parkinson's disease. Many of these substances, especially those of plant origin, are obtainable with no (or only minor) criminal involvement, and should not be underestimated, especially in the drug scene in East Germany, although only individual cases can be established with certainty [15].

6.2.10.1 Lysergic Acid Diethylamide (LSD)

The known consumption of LSD has fallen well behind that of the opiates and the various forms of cannabis on offer, although LSD abuse continues to increase. This abuse in central Europe has not yet got back to the level of the 1980s, when, for example, 20% of American High School leavers admitted having experience of LSD. Most members of the drug scene have in fact had experience with LSD. In most cases, consumption is not regular, and the proportion of experimental users is high, as would be expected. In parallel with the large increase in confirmed LSD "trips", the number of first users is also increasing [16]. The age of first use, at 17–18 years, only slightly exceeds that for cannabis.

6.2.10.2 Other Indole Derivatives

The abuse of the South American drug ololiuqui, whose main active constituent is lysergic acid amide, is rare in Europe, and this also applies to the compounds harmine and harmaline derived from lianas, e.g., of the genus *Banisteriopsis*. The hallucinogenic effects of bufotenine (5-hydroxy-*N,N*-dimethyltryptamine), which is present in the secretion of the agua toad (*Bufo marinus*), are also sought by overseas drug users only [17]. However, the abuse of psilocine (4-hydroxy-*N,N*-dimethyltryptamine) and psilocybine is of increasing significance in Germany. Both substances are present in fungi of the genus *Psilocybe*, which are found worldwide. The Central and South American species *Psilocybe exicana* and *P. cubensis* contain psilocybine in concentrations ten times greater than those present in the native European species *P. semilanceata* and *P. montana*. The products sold as "magic mushrooms" or "LSD mushrooms" are grown and harvested both in the tropics and in Germany [18].

6.2.10.3 Mescalin

The use of peyotl or "mescal buttons" is also not limited to middle America. Mescalin-containing powder is encountered more and more in Europe, where the users are primarily members of the drug scene. However, continuity of supply of mescalin is problematical. The peyote cactus *Lophophora williamsii* is also grown occasionally in Germany.

6.2.10.4 Nutmeg

Hallucinogenic effects from nutmeg can be achieved from doses of 5–30 g. Also, elemicin and myristicin, which are present in this familiar spice, can be converted by transamination reactions to the two phenylethylamines 3,4,5-trimethoxy-amphetamine (TMA) and 3-methoxy-4,5-methylenedioxyamphet-amine (MMDA), whose psychoactive effects are well known (see Fig. 6-1). Nutmeg oil is also consumed in closed wards and prisons [3, 15].

6.2.10.5 Piperidine Derivatives

Abuse of members of the nightshade family (*Solanaceae*) sometimes occurs, a practice to be very strongly discouraged for toxicological reasons. The active constituents scopolamine, atropine and hyoscyamine cause hallucinatory effects and excited states. In higher doses they have a depressant effect on the central nervous system. Deadly nightshade (belladonna), henbane and thorn apple are usually consumed by drug abusers experimentally. These plant-based drugs are acquired legally or in the drug scene [1].

6.2.10.6 Fly Agaric

Fly agaric (*Amanita muscara*), which is similar to the tropane alkaloids, is also used by drug addicts. This abuse takes place over only a short period of time.

6.2.10.7 Biperiden

The Parkinson agent biperiden causes psychotic states at high dose rates (10–15 mg/day). Abuse of this drug was at one time rare, but drug-dependent and medicament-dependent people now sometimes use it. The deliberate production of a Parkinsonoid state by the neuroleptic effects of biperiden has also been described.

6.2.11 Substances Taken by Sniffing

The problem of drug inhalation (glue sniffing) by children and young people is a phenomenon of which little is known. These substances are easily obtained everywhere at small expense, and are cheaper than cannabis or other drugs. Inhaled substances are the most important substances of first use. As well as the common solvents such as acetone (nail varnish remover), ethyl acetate, nitro-compounds used as thinners, trichloroethane (e.g. Tipp-Ex®) and various hydrocarbons, other highly volatile substances such as gasoline are inhaled at sometimes high dose rates [19]. Currently, chloroethane and refrigerant sprays of various compositions are assuming greater importance, and the inhalation of deodorant spray has already led to deaths among young people, usually by suffocation by the plastic bag pulled over the head to concentrate the volatile materials or from the neurotoxic effects of overdosing with these substances.

7. Methods of Sample Preparation for Drug Analysis

Petra Gerhards, Jörg Szigan

7.1 History of Sample Preparation

In drug screening, the biological matrix is of special importance because of its complexity. In order to detect and identify the drugs that it contains, these must be separated from the matrix. The endogenous components such as proteins and lipids, which can interfere with the analysis, must be removed. As some drugs are present in only trace amounts, a concentration stage must be included before injection (Fig. 7-1).

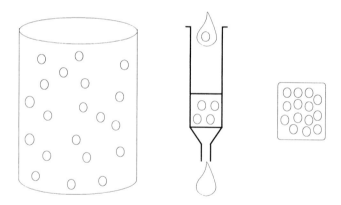

Fig. 7-1. Scheme for the concentration of drugs

For many years, liquid-liquid extraction (LLE) was used for the preparation of samples with biological matrixes, but this technique is no longer as important today. The reasons for this include:

- The large amounts of solvent
- The formation of emulsions
- The fact that automation is not possible.

Solid phase extraction (SPE) was developed during the last 10 years as an effective analytical tool for the isolation and purification of a wide range of compounds. For many applications in pharmacy and biomedicine, it is playing an increasingly important part in sample preparation both for trace concentrations and for the isolation of substances on a preparative scale [20].

SPE has some special advantages over the traditional liquid-liquid extraction technique. These include:

- Higher selectivity, as a large number of sorbents and solvents are available for the various applications
- Cleaner extracts, as the components being analyzed are selectively concentrated and can be eluted from the column (see Fig. 7-2)
- More reproducible results, as SPE is based on specific molecular interactions
- Easy handling of the columns, enabling several samples to be treated simultaneously
- The fact that automation is possible [21].

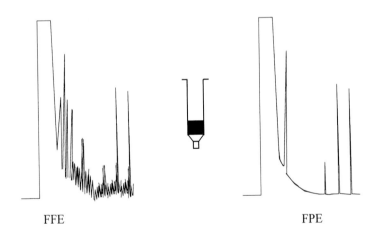

FFE   FPE

Fig. 7-2. Liquid-liquid extraction (LLE) and solid phase extraction (SPE)

7.2 The Principle of Solid Phase Extraction

Solid phase extraction is a physical extraction process that takes place between a liquid and a solid phase. The solid phase has a stronger affinity for the isolate than for the solvent in which the isolate is dissolved. When the sample solution passes through the bed of sorbent, the isolate becomes concentrated on the surface of the sorbent while the other constituents of the sample pass freely through the bed [20]. The SPE process is illustrated in Fig. 7-3.

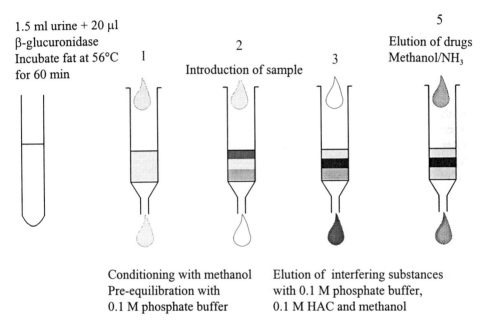

1.5 ml urine + 20 µl
β-glucuronidase
Incubate fat at 56°C   1
for 60 min

2
Introduction of sample

3

5
Elution of drugs
Methanol/NH₃

Conditioning with methanol
Pre-equilibration with
0.1 M phosphate buffer

Elution of interfering substances
with 0.1 M phosphate buffer,
0.1 M HAC and methanol

Fig. 7-3. The principle of sample preparation by solid phase extraction (SPE)

I. Column Conditioning

To ensure high and reproducible recovery, it is necessary to treat the column sorbent with a solvent. For cartridges, in which the sorbent is based on bonded silica gel, this procedure is necessary as the dry sorbent is pre-packed, and in this condition the active side chains are not available for interaction with the components. For non-polar and multi-bed phases, the sorbent must in some circumstances be preconditioned with several solvents, such as methanol followed by water or a buffer solution. The organic solvent used here has the function of solvating the bound functional groups and of activating the hydrocarbon chains. Also, organic residues are removed from the sorbent. Water or a buffer solution is then added to prepare the column for the aqueous sample.

II. Sample Introduction

After the preconditioning of the column, the prepared sample is added and then drawn through by means of a slight vacuum. A definite low flow rate is necessary in order to concentrate on the column all the various types of drugs which can be present in a biological sample. Normally, the flow rate is 1.5 ml/min.

III. Washing of the Column and Adjustment of the pH

Not only the drugs, but also many endogenous components become concentrated on the column, so that it must be washed with water or a suitable solvent in order to selectively remove the impurities. The solvent used for washing must be carefully selected so that no loss of the drugs occurs during this step. As it is not easy to find a solvent that fulfills this requirement one hundred per cent, a compromise must be found between acceptable recovery rates and an acceptable removal of the impurities. The cleaning process must be appropriate for the subsequent analysis. In some cases, the pH of the column is fixed by this step. This is necessary to give the required selectivity for the drugs, whose elution depends on the pH.

IV. Drying of the Column

If the eluent used, e.g. chloroform, is not miscible with water, it is advisable to remove any aqueous residues from the column so that this water will not cause problems in the elution of the drugs and in the GC analysis of the extract. Aqueous residues on the column can be removed by suitable organic solvents, centrifugation, or application of a vacuum.

V. Elution of the Drugs

The elution of the drugs from the column is carried out with a suitable solvent. This must be powerful, so that a small amount will completely elute the drugs. It must also be selective, so that interfering components are not eluted at the saame time [21].

7.2.1 Practical Aspects

In recent years, several mixed-bed cartridges for drug analysis have been developed. Columns of this type include Bond Eluat Certify (Varian), Clean Screen (Worldwide) and Narc-2 (Baker).

These SPE cartridges contain cation exchange groups and lipophilic groups (C-18 or C-8 groups). These enable a wide range of drugs to be retained and then selectively eluted. The drugs are eluted in two fractions:

- Basic drugs
- Acidic drugs.

The acidic drugs are eluted in the first fraction, and the basic drugs in the second. The extracts and the recovery rates, especially in the case of the benzodiazepines (flunitrazepam), are cleaner and better using the mixed-bed phase than using pure C-18 columns [21, 22].

The materials used are:
− Methanol, acetone, chloroform (Merck, HPLC quality)
− β-glucuronidase (Merck)
− Nalorphine (Sigma or Promochem/Wesel))
− Phosphate buffer (0.1 mol/l, pH 6, KH_2PO_4 analytical grade, Merck)
− Acetic acid (0.1 mol/l, pH 3)
− Ammonia/methanol (1:9 v/v), must be freshly prepared every day.

7.2.1.1 Sample Preparation for the Matrix Urine

As many drugs, e.g., benzodiazepines and opiates, are excreted as glucuronides, an enzymolysis is necessary before the SPE is performed, but, in the case of opiates, it is important that the period of incubation with β-glucuronidase should not exceed 45 min. This is to prevent 6-monoacetylmorphine from being hydrolyzed to morphine, thereby preventing the detection of possible heroin abuse. Because of the high salt level in the urine, the latter must be diluted with water or a phosphate buffer. This is especially important if ion exchange takes place first as an extraction stage. The cations of the salt compete with the drugs for the free bonding sites, leading to poorer recovery rates.

The toxicologically relevant drugs are basic or acidic, and can therefore be eluted successively from the column by using different pH values (see Fig. 7-4). This is achieved by washing the column with 0.1 M acetic acid (pH 3). With this conditioning, the acidic, neutral and slightly basic drugs exist as non-polar components, and can be eluted from the column by lipophilic solvents. The basic drugs are retained by the cation exchange groups and are then eluted by ammoniacal methanol.

Incubate 1.5 ml urine + 20 µl β-glucuronidase at 56°C for 60 min and
add 50 µl nalorphine (internal standard 100 µg/ml)
⇓
Activate the FPE cartridge with 3 ml methanol
and condition with 3 ml phosphate buffer
⇓
Mix urine with 3 ml 0.1 M phosphate buffer and pass through the cartridge
using a slight vacuum
⇓
Wash with 3 ml 0.1 M phosphate buffer
⇓
Treat with 0.5 ml 0.1 M acetic acid, pH 3, using a slight vacuum
⇓
Pass 50 µl methanol through the cartridge using a slight vacuum
⇓
Dry by applying high vacuum
⇓
Elute the acidic fraction with 4 ml acetone/chloroform (1:1)
⇓
Dry by applying high vacuum
⇓
Elute the basic fraction with 2 ml ammoniacal methanol (1%)
⇓
Evaporate the eluates separately at 40°C in a stream of N_2
⇓
Incubate the residue with acetic anhydride for 30 min at 60°C
⇓
Evaporate
⇓
Take up the residue in 50 µl methanol
⇓
Inject 1 µl into the GC for analysis

Fig. 7-4. Scheme for sample preparation in drug analysis [21, 23]

The recovery rates using Bond Eluat Certify and the sample preparation procedure described above lie between 80 and 103% [21].

In drug analysis, serum, plasma and whole blood are analyzed as well as urine. Serum and plasma, like urine, can be processed without additional treatment.

The main problem with whole blood is the presence of the red corpuscles, which can adhere to the cartridge and lead to poor reproducibility. Some drugs can bind to the red corpuscles or become entrapped inside them. It is therefore important to hemolyze the red corpuscles before injection onto the column.

The various methods of hemolysis that can be used include the addition of methanol, acetone or zinc sulfate/methanol, or the use of ultrasound. Zinc sulfate has the disadvantage that it forms zinc hydroxide with hydroxyl ions and blocks the column. Both solvents have the disadvantage that losses can occur on injection of the sample. To prevent these losses, the solvent should be evaporated off in a water bath before introducing the sample to the column. All three methods have the further disadvantage that they cause protein precipitation, causing drugs to be entrapped. The use of an ultrasound bath is therefore preferable.

7.2.1.2 Sample Preparation for the Matrix Whole Blood

The whole blood (1 ml) is placed in an ultrasound bath at ambient temperature for 15 min. Phosphoric acid (6 ml, 0.1 M, pH 6) is added, and the mixture is vortexed for 30 s and centrifuged at 1500 rpm for 15 min. This is then followed by the sample preparation procedure described above [21].

7.3 Liquid-Liquid Extraction (LLE)

In liquid-liquid extraction, Toxi-Tubes A produced by DRG Toxilab are used. Urine (5 ml) is added to the reaction liquid in the tube and mixed by shaking for 2 min. The tube is then centrifuged for 5 min at 2500 rpm. The clear supernatant is separated and evaporated. The residue is acetylated and injected into the GC. As well as the Toxi-Tubes A, Toxi-Tubes B are manufactured for acid extraction, especially for barbiturates.

SPE and LLE are compared below.

7.4 Comparison of SPE with LLE

Flunitrazepam and morphine are used here as examples for this comparison.

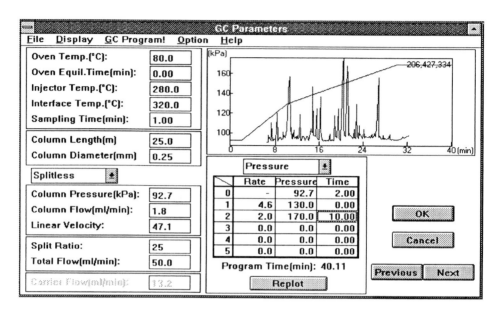

Fig. 7-5. GC/MS parameters used

After sample preparation, the samples, with the parameters shown in Fig. 7-5, were measured. The column used was a nonpolar 25 m capillary (e.g. SE-54-CB or HT-8), 0.25 mm I.D., 0.25 µm film. The column was run using a temperature program: 80°C for 2 min, 15°C/min increase to 200°C, 7°C/min increase to 295°C, 20°C/min increase to 340°C, and holding at 340°C for 6 min. The high initial temperature in the column program was used to avoid long cooling times for the gas chromatograph. This enabled a higher sample throughput per day to be achieved.

The following criteria were compared:

- Background
- Responses of the individual components
- Substance groups detected.

Figure 7-6 shows a chromatogram and the MS spectrum of flunitrazepam. This sample was prepared with a Bond-Eluat-Certify cartridge.

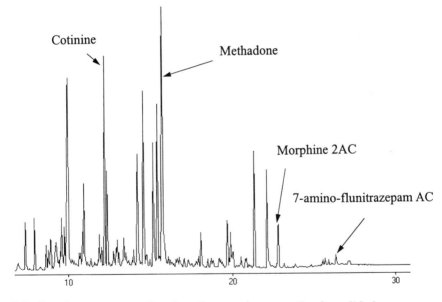

Fig. 7-6. Gas chromatogram of a urine after sample preparation by solid phase extraction (SPE). MS spectrum of 7-amino-flunitrazepam AC

The comparison between SPE and LLE shows that with LLE both acidic and neutral components such as barbiturates and ibuprofen are detected. This can be seen from Fig. 7-7.

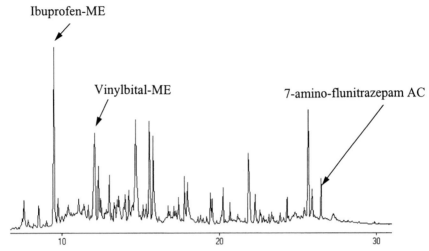

Fig. 7-7. Gas chromatogram of a urine after sample preparation by liquid-liquid extraction (LLE)

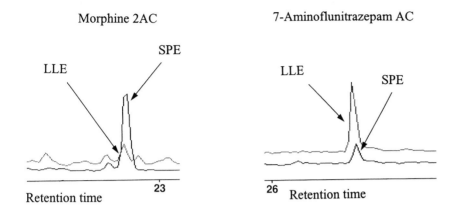

Fig. 7-8. The effect on thegas chromatogram of sample preparation of morphine and 7-amino-flunitrazepam by solid phase extraction (SPE) compared with liquid-liquid extraction (LLE)

It can also be said that the detection of morphine using LLE is poorer. Figure 7-8 shows a comparison between morphine and 7-amino-flunitrazepam prepared using SPE and LLE. With LLE, better results are obtained for 7-amino-flunitrazepam, as shown by the peak height.

In general, it can be said that SPE is a good method of purifying and concentrating urine. The advantage of LLE lies in the saving of time. Especially in emergency analysis, time is an important factor. LLE may also be used if a matrix is of such a consistency that SPE is not possible. An example of this is the analysis of stomach contents for a toxicological investigation [22].

Special Note
The sample preparation procedures described above can be used for the majority of drugs of abuse. Exceptions include cannabis, LSD and one or two others. A special preparation technique is used to determine these substances, but this is not discussed further here.

7.5 Quantitative Determination of Derivatized Drugs

7.5.1 Introduction

In some cases, the quantitative determination of an abused drug may be needed, for example, if this is prescribed by Section 24a of the German traffic law.

   Cannabis is by far the most commonly used illicit drug. Most people who abuse hard drugs such as heroin also abuse cannabis. Tetrahydrocannabinol (acid and Δ^9-metabolite) is regulated in the German traffic law (see Table 7-1), and amphetamine, opiates (morphine) and cocaine (as benzoylecgonine metabolite) are also mentioned.

7.5.2 Determination of the Detection Limits

To determine the detection limits of the system, a methanolic standard solution is first analyzed. The standard solution contains the native substance and also the d_3-labeled substance in which three hydrogen atoms are replaced by three deuterium atoms, this being used as an internal standard for the quantification of the native drug. A standard solution is measured first because the derivatization of the compounds can vary. Some analysts use MSTFA or BSTFA for derivatization. For our application, PFP (pentafluoropropionic anhydride for -OH groups and pentafluoropropanol for -COOH groups) is used. One reason for using this reagent is that it is compatible with the other derivatization reagent, acetic anhydride, which we also use. If TMS is used, there can be problems if the previous sample analyzed was derivatized with acetic anhydride, which dissolves TMS from the glass liner and the column, so that the ion source can be contaminated. Other possible methods for the derivatization of opiates are the use of propionic anhydride or methylation. Few PFP or propionic anhydride derivatives are included in the Pfleger-Maurer-Weber library. Table 7-1 shows the compounds to be determined, the molecular masses of the unlabeled and d_3-labeled compounds used and the legally required detection limits.

Table 7-1. Used *m/z* and limits

Compound	Native masses	d₃ masses	Limit by law in whole blood (ng/ml)
Amphetamine PFP	118; 190	121; 193	50
Benzoylecgonine PFP	300; 421	303; 424	50
DELTA-9-THC PFP	377; 460	380; 463	5
Morphine PFP	414; 577	417; 580	10

Figure 7-9 shows the ion sets in single-ion monitoring (SIM) for the determination of the detection limits. A bovine blood was spiked with the unlabeled and the deuterated compounds in the concentrations shown in Table 7-1 derivatized with PFP.

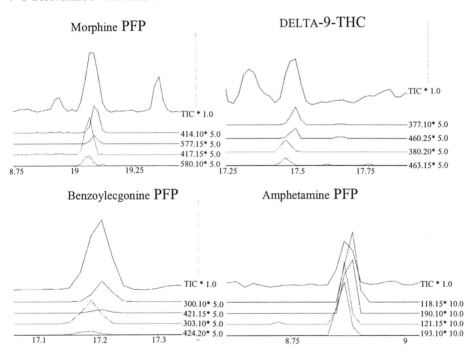

Fig. 7-9. SIM ion sets for the derivatized drugs

7.5.3 Sample Preparation and Analysis of Real Samples

Before starting the sample preparation, it is necessary to know the concentration/time curve of the compound in blood. The deamination time depends on the compound and in some cases also on the abuse behavior of the person. It is very important to know the kinetics. Cocaine is metabolized to benzoylecgonine is recommended for the determination of cocaine abuse. These kinetics are very well explained by P. X. Iten [23]. Figure 7-10 shows the curve for amphetamine as an example.

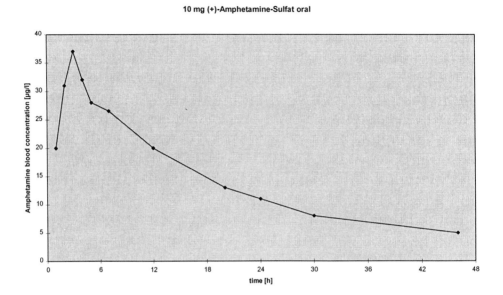

10 mg (+)-Amphetamine-Sulfat oral

Fig. 7-10. The metabolism of amphetamine in blood (page 89 in [1]).

The sample preparation is carried out as described in Chapter 7. The derivatization is performed with 70 µl PFP alcohol and 100 µl PFP anhydride. If PFP is used for the derivatization it is necessary to use water-free ethyl acetate to dissolve the residue.

For the determination of THC-COOH and also Δ9-THC, a C-18 cartridge preconditioned with methanol can be used, and the elution is done with acetonitrile. We normally use the complete sample preparation procedure for the determination of all compounds that could be present.

In some cases, sample preparation by liquid-liquid extraction (LLE) is used. With a good solid phase extraction (SPE), a concentration factor of up to 40 can be achieved. This concentration factor also means that the detection limits for the method can be 40 times better than the detection limits for methanolic standards. This is why we use SPE to generate our calibration curves for drugs in blood. A bovine blood is spiked with the standards and afterwards prepared by SPE. By generating a calibration curve from a real matrix all matrix effects are taken into consideration. Figure 7-11 shows a calibration curve for morphine-PFP.

The results and the sensitivity are much better when methanolic standards are used for the calibration curve, because then no matrix effects are taken into consideration. Using SPE as the sample preparation method, a detection limit for

THC of 100 fg can be achieved. Figure 7-12 shows the determination of 3 ng/ml Δ^9-THC in whole blood using LLE [24].

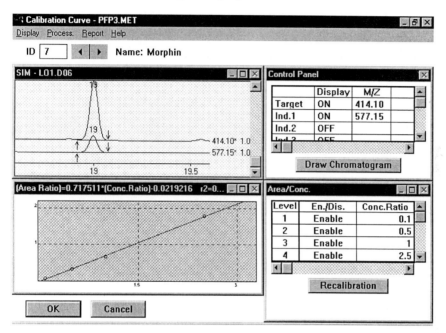

Fig. 7-11. Calibration curve for morphine

Another example of the quantitative determination of a drug is provided by LSD. Because of the very high retention index (3595) of the LSD-TMS derivative, chromatography is not easy. Well-deactivated silylated inserts are recommended. The extracts from a human specimen should be free of matrix (reextraction). For the derivatization, the extremely dry residue is taken up in BSTFA/TMCS (Fluka 15238) and derivatized for 20 min at 70°C. Instead of pure BSTFA/TMSC, a 30–50% solution in a dry solvent such as acetic acid, diethyl ether or dioxan can be used.

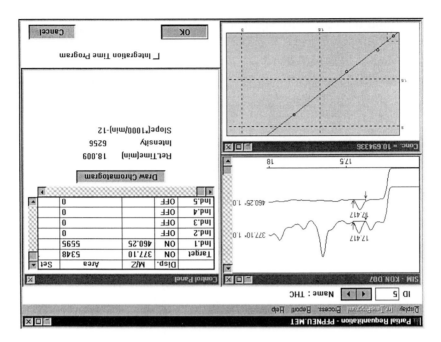

Fig. 7-12. Determination of Δ^9-THC in a real sample

For the determination of the LSD derivatives, a GC-17A in combination with a QP-5000 has been used. A J&W column type DB-5MS with ID 0.25 mm, length 25 m and film thickness 0.25 μm was used for the chromatographic separation.

In view of the high retention index of LSD/TMS, the use of a shorter column with a thinner film should be considered. Using single-ion monitoring (SIM) with GC/MS, the separation can be done on a shorter column without endangering the quality of the results.

Figure 7-13 shows the multi-ion chromatogram generated from the full scan and the library search result of a 20-ng LSD standard which was derivatized with BSTFA.

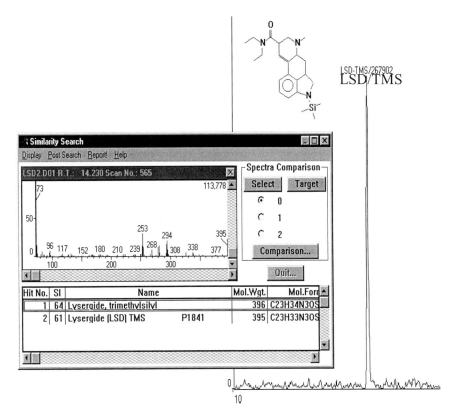

Fig. 7-13. Scan mode chromatogram and library search result of 20 ng/ml LSD/TMS
(absolute) measured in full scan

It should be noted that derivatization with BSTFA requires a completely water-free atmosphere. In the presence of small amounts of water (atmospheric moisture levels), derivatization can be disturbed, and loss of substance can occur. To derivatize the sample, the vial is put on the hot injector for around 45 min. The sample should then be analyzed immediately, as the peak area decreases with increasing time.

Figure 7-14 shows the extract of a urine sample spiked with LSD/TMS and LSD-d_3/TMS. The samples were measured in SIM, in which two significant masses for LSD and the internal standard LSD-d_3 were chosen. When analyzing LSD, the samples must be stored in brown glass vials or protected from light in order to prevent loss of the compound due to reactive breakdown.

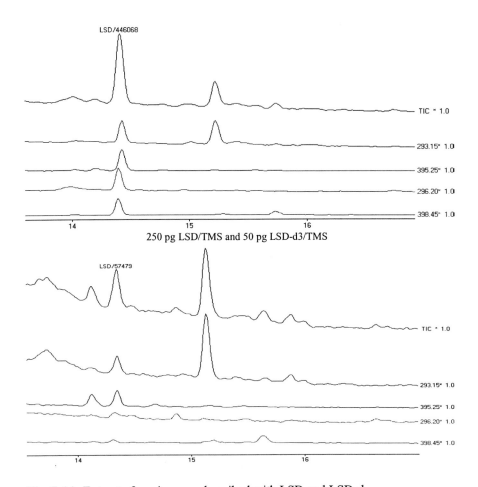

Fig. 7-14. Extract of a urine sample spiked with LSD and LSD-d$_3$

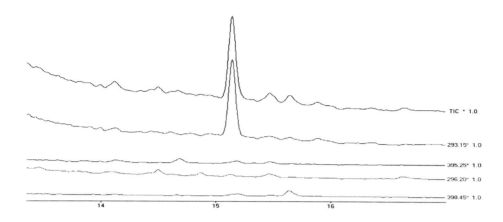

TIC · 1.0

293.15· 1.0

395.25· 1.0

296.20· 1.0

390.45· 1.0

14   15   16

Fig. 7-15. Extract from a non-spiked urine

Figure 7-15 shows the extract after solid-phase extraction (SPE) from a non-spiked urine. This experiment was carried out to investigate the influence of the matrix.

SPE shows some advantages over LLE:

• Purity of the extracts
• Concentration of the compound.

Acknowledgement. We would like to thank Dr. K. Besserer of the Institute of Forensic Medicine, University of Tübingen, for his support for the LSD application.

8. Drug Screening from Urine by GC/MS

Ulrich Bons, Jürgen Sawazki

8.1 General Aspects of Drug Screening from Urine

The object of drug screening is to detect drugs of abuse or to demonstrate that they have not been taken (see Chapter 6). Drug screening has continually increased in importance over recent years in parallel with the increase in drug consumption and the developments in analytical equipment, no end to these trends being currently in sight. Thus, this technique can be important in detoxification therapy for drug abusers, or alternatively could serve to rule out drug-induced psychoses and support a psychiatric diagnosis. The results of drug screening are also important in the investigation of possible criminal offenses or illegal drug dealing. Also, as an accompaniment to methadone substitution therapy for drug abusers, the legal authorities require the routine use of approved diagnostic laboratory procedures. This analysis must be able to detect the co-consumption of all those substances the abuse of which compromises the object of the substitution procedure [26]. Here, the most important aim is to prevent any fatalities which could occur if patients undergoing substitution therapy consume other drugs and medicaments in addition to methadone, behavior which could lead to supra-additive pharmacological effects.

The analytical methods used must be sensitive to the broad spectrum of relevant substances from antidepressants to cerebral stimulants of the amine type, bearing in mind the variations in abuse behavior that must continually be kept under observation, so that the method remains credible and the patient can be treated in the best possible manner.

The authors attempt to describe the potential applications and limitations of GC/MS as a screening method for the rapid and reliable determination of drugs of abuse and to compare it with alternative viable procedures.

8.1.1   Differentiation between Drug Screening and
Other Screening Procedures

In addition to the above-described methods for determining drugs of abuse, screenings are also carried out to answer toxicological questions or in the area of doping control. Toxicological screening is used to detect organic and inorganic compounds in a biological matrix. In this connection, it should be noted that ethical drugs and drugs of abuse are the second most important cause of poisoning after ethanol intoxication [27]. GC/MS is a combination of techniques that is very well adapted to the identification of most organic compounds, as no other analytical equipment can determine a larger number of such substances with equal sensitivity and specificity [28]. The method described in this Chapter is also applicable to the investigation of intoxication by ethical drugs and drugs of abuse, and leads to concrete results in a short time, which is extremely important in clinically relevant investigations. Doping control mainly involves the elimination of the possibility that improvements in sporting achievements are the result of, e.g., the use of anabolic substances such as hormones or other medicaments. GC/MS is an important aid in this screening process, and the methods described below are used for the investigation of certain aspects of the problem.

8.1.2 Sample Material

The use of blood as a sample material for drug screening has hitherto been of minor importance. However, blood or serum is now being increasingly used in the detection of drugs of abuse, especially in connection with traffic offenses, as only precise quantitative results can enable conclusions to be drawn concerning the competence to drive of the persons concerned [29]. Interlaboratory investigations on this subject are described by the German association for toxicological and forensic chemistry (Gesellschaft für Toxikologische und Forensische Chemie, GTFCh). The wide variations in the concentrations in urine of drugs of abuse and the abundance of metabolites of a given substance (e.g. phenothiazine) sometimes found in urine means that blood presents itself as the most suitable starting material for the reliable determination of concentrations. Moreover, blood contains fewer interfering substances than urine. The investigation of hair is a special case of drug screening in which the length of the hair enables drug consumption over long periods of time to be investigated [30]. The exact earmarking and identification of the sample material is especially important in hair analysis [31]. The use of the GC/MS system is especially suitable for the analysis of drugs in hair because of its very high sensitivity. However, the sample preparation described in this book has not been tested with hair as the sample material.

For most questions involving drug screenings, the test material routinely used is urine. This material is as a rule free of problems, does not involve invasive procedures and is obtainable in large amounts, and the concentrations of the medicaments and drugs are always appreciably higher in urine than in blood. Disadvantages include the above-mentioned physiologically caused concentration fluctuations of abused substances in urine, so that false interpretations of quantitative statements about the drug content of a urine sample can very easily be made. This is especially likely if a patient has consumed a drug with a long elimination half-life and is tested daily. In normal circumstances, the first morning urine sample of a drug user contains the highest concentration of the drug, but there are many possibilities for the drug content to be influenced, whether deliberately or not. For example, if a patient takes in a large amount of liquid before the urine sample is taken for a drug screening, the drug concentration in the urine can be considerably reduced, and can sometimes briefly fall below the limit of determination of the screening method chosen, causing a "detection gap". Measurement of the creatinine content can give an indication of any manipulation of the urine by dilution, and can be recommended as a credibility test. In many laboratories, urine samples whose creatinine content falls below 30 mg/dl (1760 µmol) are regarded as having been manipulated [32]. The pH of the urine also affects the drug concentration. Normal values of urine pH are in the range 5–8, depending on the metabolic situation. Most relevant substances in drug screening are bases, so that concentrations of these in urine at pH 5 are higher than those at pH 8, as in slightly acid media these substances are partly present in protonated form, so that resorption is hindered. The pH of the urine can also be influenced by the oral ingestion of large amounts of ammonium chloride pastilles, ascorbic acid (pH reduction) or sodium carbonate (pH increase).

8.1.3 Possibilities for Manipulation of Samples

Patients who believe that samples taken from them will show positive analytical results frequently attempt to interfere with the analysis by adding various substances to the urine in order to hinder the detection of the drugs. Immunological methods can sometimes even give false positive results after manipulation of the urine, which, in the absence of confirmatory analysis, can place the credibility of the chosen method in doubt. Drug screening by GC/MS is considerably less susceptible to manipulation than are immunological methods, although the addition of a surfactant to the urine can cause considerable problems in sample preparation for GC/MS analysis by interfering with the extraction process. Control of the pH of the urine is not a sufficient precaution against the addition of surfactants. However, a large addition of surfactant can usually be detected from the cloudiness produced in the sample. The best protection against unpleasant surprises caused by manipulation of the urine must always be the consistent monitoring of the method of obtaining the sample.

8.1.4 Immunological Methods of Analysis

There are two principal methods of drug screening, namely the immunological and chromatographic methods. The immunological methods detect the presence of drugs in urine by an antigen-antibody reaction, which is measured in various ways. Immunological methods give a result within a few minutes and, for a first test, demand a lower level of expertise and experience from the analyst than do chromatographic methods. However, extensive knowledge of the cross-reactivities of the methods used and the metabolism of the relevant substances are essential preconditions for the responsible application of immunological tests and interpretation of the results in drug screening.

In this context, the cross-reactivity characterizes whether and to what extent individual antigens of various structures are detected by particular antibodies. Even those structures not associated with drugs can show a cross-reactivity to antibodies in drug tests. A very comprehensive investigation and documentation of the cross-reactivities of the immunological tests used is therefore as important for the analyst as an adequate sensitivity of detection of the relevant medicaments and drugs and their metabolites. Whereas many manufacturers' tables of cross-reactivities of immunological reagents towards important medicaments are very comprehensive, the results for the metabolites produced from these substances are often incomplete. This subject is discussed in more detail below, where the various types of drugs are dealt with individually.

In recent years, rapid tests for immunological drug screening have become more available from a large number of companies, many of which are simply marketing organizations. In contrast to the classical immunological methods, these rapid tests can be used without sophisticated equipment and can provide immediate results on the spot to law enforcement officials, drug treatment experts, care workers, or general practitioners dealing with methadone substitution therapy. Unfortunately, the documentation of the cross-reactivities of such tests is often sparse, which leaves open many questions concerning the ability of the method to detect relevant substances or undesired cross-reactivities. It is then left to the user to first of all carry out a number of comprehensive and hence expensive investigations to validate the methods prior to using the material in a responsible manner. However, this situation does not deter the user, who may be an inexperienced analyst and poorly informed about the problems described above, from using such tests. The risk of false interpretation of the "information" that these rapid tests yield is correspondingly high. The need for confirming such information, including negative information, by chromatographic procedures can therefore not be overemphasized.

8.1.4.1 Immunological Tests on Single Substances and Groups of Substances

Reagents produced especially for the determination of particular drugs such as cannabis (11-nor-Δ^9-tetrahydrocannabinol-9-carboxylic acid), benzoylecgonine (a cocaine metabolite) or methadone are characterized by their high sensitivity and specificity. The often high sensitivity of the tests is a function of the targeting of the antibody, usually towards a single very characteristic antigen structure. This leads, for example, to the situation that an easily obtainable commercial test is able to detect the cocaine metabolite benzoylecgonine with great sensitivity (300 ng/ml), although a urine sample from a drug-free patient gives a negative result even after addition of appreciable amounts of cocaine (see Section 8.3.6). Compared with group tests, the high specificity of the antibody gives a smaller risk of undesired reactions with other antigen structures, reducing the danger of false positive results.

In immunological tests on amphetamines, barbiturates, benzodiazepines and opiates, results that apply only to groups of substances can be obtained, where each group comprises a large number of relevant substances. The antibody, developed using one or more of the substances often present in urine which are members of the group concerned, has variable reaction sensitivity towards the individual drugs and metabolites of this group of substances. The greater the structural differences between the substances being measured and the calibration substances, the more can the cross-reactivities differ. This means that the members of a large group of substances and their metabolites can have very different cross-reactivities. The individual relevant substances are therefore detected with varying sensitivity. Thus, in many cases, consumption of the low therapeutic doses of lorazepam (0.5–2 mg/day) commonly prescribed today is not detectable because of the low cross-reactivity with the known benzodiazepine reagents. The low affinity of lorazepam to benzodiazepine antibodies thus leads to a false negative result in many cases. The cross-reactivities of the immunological group tests of different manufacturers towards antigen structures present in urine can differ considerably. The discussion of the relevant groups of drugs of abuse (see Section 8.3) therefore also includes the current data from the manufacturers on the cross-reactivities of the most important of these substances towards immunological reagents. A reaction of a non-relevant substance with an antibody is also theoretically possible. This would give a false positive result, with potentially grave consequences for the patient and for the credibility of the analysis. To avoid such a situation, all positive results of immunological tests are confirmed by some other method. It should also be noted that immunological reagents are not available for all relevant substances or substance groups. Antihistamines, biperiden, chlormethiazole, tilidine and tramadol cannot (yet) be determined by immunological techniques.

8.1.4.2 Principles of Current Immunological Tests (Selected)

CEDIA (Boehringer Mannheim)

The abbreviation CEDIA stands for homogeneous cloned enzyme donor immuno-assay. The CEDIA test is based on the the enzyme β-galactosidase, which is split by genetic technology into two inactive fragments, an enzyme acceptor (EA) and an enzyme donor (ED). On addition of the EA to a solution containing ED, the fragments recombine spontaneously to form the intact enzyme. The ED is coupled with the antigen structure (drug). In the sample, the free antigen structure (drug) competes with the ED antigen structure for the antibody. An ED-antigen-antibody complex is no longer able to recombine with the added EA to form an intact enzyme. The number of recombined enzymes formed in the sample correlates with the amount of free antigen structures (drugs) originally contained in the sample. The extent of the substrate reaction of the intact enzyme is measured spectrophotometrically at 660 nm [33].

EMIT

Enzyme-linked multiplied immuno-assay (EMIT) is also a homogeneous enzyme immuno-assay. A definite amount of an antigen marked with glucose-6-phosphate dehydrogenase is added to the urine sample. This antigen competes with the analyte (drug) contained in the urine for the bonding sites (which are available in limited amounts) of the antibodies. Excess glucose-6 phosphate dehydrogenase, whose activity is not greatly weakened by the immune reaction, now oxidizes glucose-6 phosphate and reduces nicotinic acid amide adenine dinucleotide (NAD) to the hydrogenated compound NADH. The NADH formed is determined spectrophotometrically at 340 nm. The measured UV emission is directly proportional to the concentration of the antigen (drug) contained in the urine.

FPIA

The measurement principle of fluorescence polarization immuno-assay (FPIA) is based on the measurement of the change in polarization of emitted light. Fluorescein, on excitation by short-wave polarized light, emits green fluorescence with rotation of the plane of polarization. An antigen structure (drug) bonded to fluorescein leads to a lower degree of polarization than a fluorescein-antigen-antibody complex. Antigen structures (drugs) from the sample and fluorescein-antigen compounds compete for binding to the antibody. The resulting decrease in polarization is directly proportional to the concentration of the drug present in the sample [34]. This measurement principle is able to detect very low concentrations of a drug, as even small decreases in polarization can easily be measured. The user could arrange for the threshold value (cut-off) to be reduced in order to further improve the sensitivity of the method. However, because of the large number of possible sources of interference, a reduction in the cut-off value is not advisable, as this can lead to an increase in the number of false positive results.

Triage™ 8 (Merck)

Whereas the immunological tests described above can only be used in conjunction with a photometer suitable for the measurement principle concerned, the Triage test simply requires a test cassette which is approximately the size of a credit card. This makes it possible to conduct the test in the presence of the patient. In contrast to the photometric test methods, which express the analytical results in numerical form, a positive result is indicated by the development of a color in a result zone. With the aid of the test cassette, a urine sample can be tested for up to eight substances or groups of substances. In the testing system, gold-bonded antigens compete with the drug for the bonding sites on the antibody. In the result zone, monoclonal antibodies are bonded to a membrane. If no drug is present in the sample, all the gold-bonded antigens are bonded by free antibodies. No further reaction is then possible in the result zone. Only in the presence of the drug in the sample do free antigen structures come into contact with the monoclonal antibodies of the result zone and cause a color change due to the antigen-antibody reaction [35]. A disadvantage is that the extent of drug consumption is more difficult to estimate from this test than from results determined photometrically.

8.1.5 Chromatographic Methods

Of all the chromatographic methods, high-performance liquid chromatography with diode array detection (HPLC/DAD) and gas chromatography with mass spectrometric detection (GC/MS) are currently the most important. Both methods detect single substances and their metabolites, which distinguishes them from the immunological group tests. Moreover, thin-layer chromatographic methods give the possibility of confirming positive results obtained using immunological methods. Often, if medicaments or drugs of abuse are taken, several metabolites of these substances are excreted, and, even if concentrations in the urine are low, the various peaks from the same starting substance may be identified, increasing the certainty of the analytical results.

In addition, the analyst can obtain information about accompanying medication, which can be helpful, at least as a credibility test. Against the advantage of the more detailed and abundant information afforded by chromatographic analysis can be set the disadvantage of the greater amount of time and higher level of expertise required of the investigating personnel compared with immunological methods. Therefore, in our laboratory, chromatographic methods are not used if the same information can be obtained more quickly from an immunological test. The detection of cannabis abuse is an example of this.

Also, the use of chromatographic methods does not exclude the principle of confirming a positive result by using a completely different method, as no method is completely error-free.

8.1.5.1 Gas Chromatography with Mass Spectrometric Detection

The use of GC/MS for the determination of medicaments and drugs of abuse must be regarded as the "gold standard", as no other method enables so many substances to be determined with certainty and sensitivity [36]. As early as 1985, the GC/MS system had been put forward by the Deutsche Forschungsgemeinschaft as the most informative method used in the determination of addictive substances [27]. However, as the costs of this method of analysis were very high at that time, its use was reserved for special investigations. Even now, GC/MS is often only used in the field of forensic medicine and for confirming positive screening results. This limitation to such a small application area is hard to understand, as the modern quadrupole mass spectrometer used in drug screening provides mass spectra with satisfactory resolution, while costing only a fraction of the cost of the equipment used ten years ago. The total costs of GC/MS analysis have also fallen considerably as a result of the use of powerful computers, which now enable mass spectra to be compared in a fraction of the time needed ten years ago, when this was a manual task performed by referring to books.

The basic equipment of the analyst who is interested in drug screening from urine by GC/MS must include comprehensive data collections. To answer questions about mass spectra of drugs of abuse or their metabolites, derivatized or underivatized, and their retention indices, a number of libraries of spectra are commercially available. These can now be obtained from many equipment manufacturers and are linked to the equipment software as a standard facility. Information on the analytical methods to be used and their limits of detection or the metabolism and pharmacology of the relevant substances can be obtained from the reference books or information sheets provided by the pharmaceutical manufacturers.

In this connection, it is well known that the limits of determination of individual substances for these special investigations can be improved by optimizing a method to suit a particular problem, e.g., by the choice of certain molecular fragments (SIM). However, bearing in mind the complexity of the relevant substances, more attention should be paid to the question of how many substances or accompanying medicaments, after consumption in concentrations associated with therapy or abuse, can be determined with sufficient sensitivity in the urine using a GC/MS method of general application. The limits of detection of important substances are discussed in Section 8.2.1.5. It should also be pointed out that in GC/MS analysis the quality of the analytical results is extremely dependent on the capabilities of the people who evaluate them.

8.1.5.2 High Performance Liquid Chromatography with Diode Array Detection (HPLC/DAD)

An important method alongside GC/MS is HPLC/DAD, which is used in the analysis of addictive drugs. HPLC operates at lower temperatures than those used in gas chromatography, so that it has the advantage of being able to analyze for thermolabile substances such as the phenothiazines, which can be detected unchanged. Derivatization steps, which are necessary in the field of GC/MS analysis to achieve sensitive detection of compounds containing hydroxyl or carboxyl groups, are not necessary with this method. Thus, even individual substances and metabolites of the important benzodiazepine group can be sensitively detected without great sophistication [37]. The identification of the substances by the diode array detector is carried out using photometric measurements at wavelengths in the range 190–380 nm. Necessary preconditions for this method of analysis are therefore absorption in this range and a sufficiently high molar extinction coefficient such that a substance in low concentrations will be detected. Unlike the UV detector, the diode array detector provides important information about an unknown peak, namely its UV spectrum. The diode array detector is therefore preferable to the UV detector for drug screening in spite of its lower sensitivity. As, in some cases, many different substances can often have similar UV spectra, stable retention times are very important for the reliable identification of a consumed substance by the HPLC/DAD method, although the assignment of a peak to a substance based simply on knowledge of the retention time would be irresponsible, bearing in mind the large number of possible substances in urine.

8.2 GC/MS Methods of Drug Screening from Urine

Before describing GC/MS methods suitable for drug screening from urine, it should first be emphasized that all screening methods are compromises between detection sensitivity on the one hand and the detection of as many medicaments and drugs of abuse as possible on the other. By use of the SIM mode, in which individual significant masses are scanned, the sensitivity with respect to the TIC (see Section 4.4.4) can be considerably improved. Interfering background masses originating from matrix components, column bleeding or accompanying therapeutic medication are suppressed, and only the characteristic masses of the target substance are scanned. However, against the advantage of sensitivity must in this case be set the disadvantage of the high selectivity, which excludes all other substances from the analysis. Therefore, the SIM mode is not particularly suitable for a drug screening test. The situation is different in the United States, where the NIDA (see Section 8.2.1.5) prescribes the analysis of certain substances only, in contrast to the German law. Here it is quite logical to the use SIM, which also enables the low concentrations of drugs in serum to be determined.

As the time required for the various methods of analysis to produce a result varies, the urgency of the investigation may be another (though usually secondary) factor to consider when choosing a particular method. In consequence, a screening method cannot include all relevant substances with equal sensitivity, nor can a highly specific method detect all substances at the limit of detection. The choice of the method of sample preparation also has an influence on the sensitivity of the analysis. In the methods of drug detection described below, both the advantages and the disadvantages of these methods in comparison to other methods are described.

8.2.1 General Aspects of Screening Methods

8.2.1.1 Silanized Insert and Silanized Glass Wool

The sample is introduced via the septum of the injector into a small glass tube (the insert). After 100 injections, the septum should be changed to prevent air from passing through the pores formed in it by the needle. One end of the capillary column projects into the insert, which contains some glass wool, to prevent the capillary column from being soiled and to prolong its useful life. The glass wool also serves to give better vaporization and distribution of the sample. The medicaments and drugs or their metabolites in the urine often contain hydroxyl, amino, or carboxyl functional groups. These reactive groups can react intensively with each other in the injector or with components of the column film, which considerably reduces the sensitivity with which the equipment can separate and detect these substances. Silanization of the insert and the glass wool very effectively shields these groups on the surface of these components of the injector. Underivatized substances such as amphetamine, morphine or some benzodiazepines therefore reach the column largely unaffected and can be detected with good sensitivity. After 100 measurements, or at least after 1 week, the insert and the glass wool are replaced. Figure 8-1 demonstrates, with the aid of a standard mixture, the increase in the sensitivity of detection following the fitting of a silanized insert. Both silanized inserts and silanized glass wool are commercially available.

For reasons of cost, used inserts are subjected to the following treatment in the author's laboratory to enable them to be re-used:

An insert is cleaned with chromic acid, dried, and then

- Degreased in acetone/n-hexane (1:1) and dried in nitrogen,
- Silanized for 2 min in dimethyldichlorosilane at room temperature in a closed vessel in an ultrasonic bath.
  Warning: very corrosive and highly flammable liquid!!
- The silanization reagent is allowed to drain away well. The insert is then dried with nitrogen if necessary and placed in an ultrasonic bath for 10 min.
- The insert is dried with nitrogen.
- It is then immediately fitted or stored under a nitrogen atmosphere.

All operations must be carried out at subatmospheric pressure. It must be carefully noted that dimethyldichlorosilane is very poisonous and highly flammable. The insert must on no account be touched by hand; tweezers must be used.

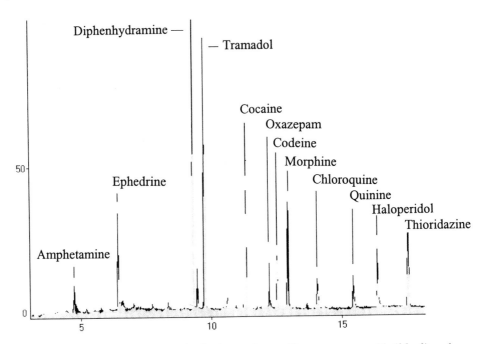

Fig. 8-1. Chromatogram of a standard mixture for quality assurance, with (*black*) and without (*gray*) silanization of insert and glass wool

8.2.1.2 Chromatographic Columns

Nonpolar capillary columns in which the stationary phase contains methyl groups as well as small proportions of phenyl and vinyl groups are suitable for drug screening. Columns of this type are available from several manufacturers. The chemically bonded phases are thermally stable up to at least 300°C. The useful lifetime of a column for drug screening depends on the purity of the sample and the quality of the sample preparation. In the author's laboratory, 800 tests of satisfactory quality can be carried out with one column. If the column is used beyond this number of tests, many substances can no longer be determined with acceptable sensitivity. Low sensitivity can also be caused by damage to the stationary phase at the end of a capillary. A better measurement result can be obtained in many cases by shortening the column at both ends.

8.2.1.3 Temperature-Pressure Programme

The optimum column temperature at the time of injection should be 20–30°C below the bp of the solvent used. For methanol (bp 65°C), the optimum column temperature should therefore be ca. 40°C. However, as cooling the column to this low temperature between the two measurements is too time-consuming, optimum focusing of the sample on the column is omitted and the analysis is begun at a column temperature of 60°C. This saving in time of ca. 10–15 min per analysis enables the frequency of sample testing to be increased. To focus the sample on the column, the temperature is held at 60°C for a period of 90 s. The oven is then heated to 160°C in 4 min. The solvent peak is suppressed as the filament is switched on only after 2.5 min. The measured signal appears after 3 min. To improve the separation of the components, the heating rate is decreased to 10°C/min (method A, see Section 8.2.2) or 5°C/min (method B, see Section 8.2.3) after 160°C has been reached. As in the latter part of the chromatogram no components that cause separation problems are eluted, a higher heating rate of 15°C/min (method A) or 20°C/min (method B) is used.

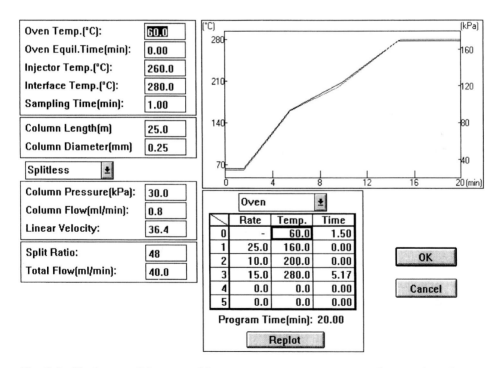

Fig. 8-2. Hard copy of the menu-driven temperature-pressure control system based on the CLASS 5000® software

The final temperature of 280°C is reached after 5.17 min (A) or 6 min (B), so that all components of the sample are eluted before the next injection. The pressure program, which is set up after the temperature program, is arranged so that the linear velocity of the carrier gas remains constant at 32.4 cm/s (method A) or 27.5 cm/s (method B) during the total duration of the analysis. A pressure program that fits the temperature program in this way is easily set up with the aid of software (see Fig. 8-2).

8.2.1.4 Internal Standard

In quantitative chromatographic analysis (e.g. monitoring antiepileptic drugs), it is standard practice to add an internal standard, as this gives a marked improvement in the quality of the result. However, use of an internal standard for drug screening from urine is not yet universal. From the point of view of the precision and accuracy of an analysis, an internal standard is not essential, as it is not possible to draw firm conclusions from concentrations of drugs found in the urine (see Section 8.1.2). However, from the point of view of quality assurance, the use of an internal standard is appropriate, as with its aid the effectiveness of the sample preparation and the success of the derivatization step, if any, can be assessed (see Chapter 13). The choice of a suitable internal standard presents problems, as the following criteria should be fulfilled:

The internal standard
● must not be present as an excretory product in the urine,
● should not interfere with the determination of substances of interest,
● should enable the quality of the derivatization to be checked,
● should react to the condition of the silanized insert and the column with sensitivity similar to that associated with the substances of interest, and
● should simulate the properties of as many substances of interest as possible on sample preparation.

It is not possible to fulfill the last requirement completely because of the large number and heterogeneity of the substances that can conceivably occur in drug screening. The substance recommended in Chapter 13, nalorphine, comes very close to being completely satifactory. This opiate antagonist does not normally occur in the urine of a drug patient, represents the properties of the opiates, and enables the quality of the derivatization to be tested. As can be seen from Fig. 8-3, nalorphine is eluted after the opiates of interest. The quantitative evaluation of peaks is therefore seldom hindered by nalorphine.

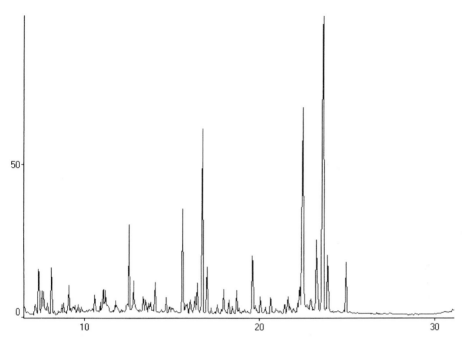

Fig. 8-3. Chromatogram of a urine with added internal standard derivatized with acetic
anhydride.
1 = Cotinine; 2 = Caffeine; 3 = Methadone-M. (EDDP); 4 = Oxazepam art.;
5 = Methadone; 6 = Nordiazepam; 7 = 6-Monoacetylmorphine; 8 = Heroin;
9 = Nalorphine-AC; 10 = Nalorphine-2AC

8.2.1.5 Detection Sensitivity

It is important for the analyst to have knowledge of the sensitivity of a method if
it is to be chosen to solve a particular analytical problem.

For determining medicaments and drugs in urine, the sensitivity of an
analytical method is usually adequate if these substances can be detected at low
consumption rates such as occur in low-dose dependence on benzodiazepines
etc. In requests for analyses, which mainly originate from psychiatric clinics,
high sensitivity methods are therefore insisted upon, as shown, for example, by
the threshold values (cut-offs) for drug screening from urine specified by the
United States National Institute of Drug Abuse (NIDA). This institute has issued
the cut-off values listed in Table 8-1 as the limits for positive drug screening
[32], which are approximated to by most immunological reagents. However, in
group tests, these cut-off values do not give any indication of the detection

sensitivity of other unlisted members of the group of substances, as the cross-reactivities of substances of a particular substance group can show very great variation. This is discussed in more detail in Section 8.3.

Table 8-1. NIDA limits of detection for drug screening from urine [µg/ml]

Substance/substance group	Compound	Cut-off value
Amphetamine	D-Amphetamine	1.0
Cannabis	11-Nor-Δ^9-THC-9-carboxylic acid	0.05
Cocaine	Benzoylecgonine	0.3
Opiates	Morphine	0.3
Phencyclidine	Phencyclidine	0.025

The methods of analysis chosen for confirmatory investigations should be more sensitive than the primary method. For example, if the positive result of an immunological group test is to be confirmed, the exact nature of the substance to be determined accurately is not yet known. Even a mixture of substances could have given the positive result. In this case a chromatographic confirmation of the result is only possible if the detection sensitivities of the methods for the substances of the group test lie well below the cut-off values of the immunological method used.

8.2.1.6 Method Validation

Before samples from patients are analyzed, every new method must be validated. In the validation investigations, the limits of capability of the method must be established and the weak points recognized. When setting up a method to be used for measuring a large number of substances, e.g. the analysis of medicaments and drugs from urine, it is not immediately possible to check the detection sensitivity of the method for all relevant substances and their metabolites. In this case, it is advisable to limit the checking to very important and analytically problematical substances. However, the determination of exact limits of detection for these substances, which would often be possible under optimal experimental conditions, is not the most important factor, as the condition of the chromatographic column, the insert and the ion source, for example, has a great influence on detection sensitivity. The data obtained should therefore only serve to enable the user to get an approximate idea of the sort of concentrations at which the substances in the urine can be determined with routinely available GC/MS equipment. Table 8-2 gives an overview of the sensitivity of detection of the methods described in Section 8.2.2.

Table 8-2. Limits of detection of drugs from urine after sample preparation

Substance		Concentration of the substance in ng/ml urine	
		Not derivatized	Acetylated
I	Amphetamine	300	50
	Ephedrine	300	50
II	3-Hydroxybromazepam	600	300
	Desmethyldiazepam	1	1
	7-Amonoflunitrazepam	1	1
	Oxazepam	1000	10
III	Benzoylecgonine	600	10 (PFP)[a]
	Cocaine	50	–
IV	Codeine	10	10
	DHC	10	10
	EDDP	10	–
	Methadone	10	–
	Morphine	300	10

[a] Pentafluoropropionic anhydride derivative

To determine the sensitivity of detection, mixtures of drugs and their metabolites (I–IV) were prepared. We deliberately mainly chose substances which can interact with the materials of the injector and the column film because of their reactive groups and which therefore cannot be analyzed by GC/MS totally without problems. All these substances are possible components of a urine sample.

First, the methanolic solutions were injected in underivatized form in increasing concentrations (0.5 ng/ml – 10 μg/ml) and measured by method A. In the next step, we added 3% of the methanolic solutions to a drug-free urine. After solid-phase extraction and concentration of the eluate by evaporation (see Section 8.1.2.10), 1 μl of the solution was injected. The concentrations which still enabled a substance to be identified immediately and the three next lower concentrations were analyzed after acetylation (see below) with the gradients of method B in order to determine the limits of detection after derivatization also.

The calculation of the retention indices for all important substances is also an important and necessary task. Retention index tables were produced in 1958 by Kováts using *n*-alkanes as standards to enable gas chromatographic retention time results obtained within a laboratory and by a number of laboratories to be compared [38]. However, according to Kováts, retention indices are defined only for isothermal operation [39] and therefore cannot be used for a temperature program. Nevertheless, retention index tables can also be used for our methods. If a mixture of ca. ten substances which cover as broad a retention time spectrum as possible is injected, the information contained in a retention index table can be used for this work. The retention index of a substance is calculated from its retention time and the known retention indices of two substances that were eluted before and after the substance under investigation (see Fig. 8-4). The information contained in the retention index can be used directly for the confirmation of results (see Section 8.2.1.6) when interpretating uncharacteristic mass spectra obtained in drug screening. In Table 8-3, the

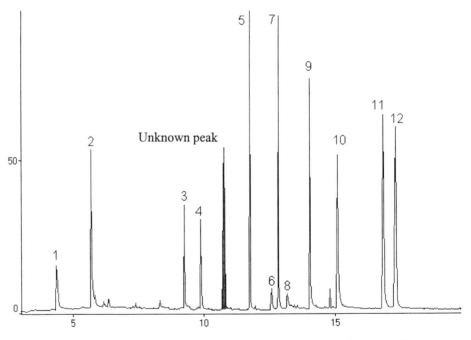

Fig. 8-4. Mixture of substances for the calculation of retention indices of unknown Peaks

Number	Retention time RT (s)	Substance	Retention index (RI)
01	258	Amphetamine	1160
02	342	Ephedrine	1375
03	552	Diphenhydramine	1870
04	594	Tramadol	1973
	642	**Unknown peak**	????
05	702	Cocaine	2200
06	750	Oxazepam	2320
07	768	Codeine	2375
08	792	Morphine	2455
09	840	Chloroquine	2575
10	906	Quinine	2800
11	1008	Bromperidol	3050
12	1038	Thioridazine	3125

contained in the retention index can be used directly for the confirmation of results (see Section 8.2.1.6) when interpretating uncharacteristic mass spectra obtained in drug screening. In Table 8-3, the retention indices for several substances obtained from the literature [38] are compared with indices calculated with the aid of comparable substances.

The retention index of the unknown peak is calculated with the aid of the neighboring substances:

$$RT_{unknown} - RT_{tramadol} = 48 \text{ s}$$

$$RI/s = \frac{(RI_{cocaine} - RI_{tramadol})}{(RT_{cocaine} - RT_{tramadol})} = \frac{227}{108 \text{ s}} = 2.10 \text{ RI/s}$$

$$RI_{unknown} = RI_{tramadol} + 48 \times 2.1 = 1973 + 100.8 = 2073.8$$

Reference to the library of reference spectra shows that the unknown peak could be metoprolol. The retention index for metoprolol is given in the literature as 2080 [38].

Table 8-3. Comparison of some retention indices taken from the literature with retention indices calculated by method A

Substance	Measured retention time	Retention index from the literature	Calculated retention index
MDA	6.58	1495	1474
MDE	6.91	1560	1555
MDMA	6.65	1790	1510
Metamizol	10.33	1995	1998
Methadone	11.36	2160	2180
Methadone M (EDDP)	10.60	2040	2061
Methylecgonine	6.31	1465	1465
Methylephedrine	6.12	1430	1435
Metoclopramide	13.69	2610	2530
Metoprolol	10.70	2080	2074

As can be seen from Table 8-3, the calculated retention indices mostly correspond well with the values given in the literature. Discrepancies greater than 50 retention index units are rare. We could not confirm the literature value of the retention index for MDMA with our measurements. Every laboratory when validating methods for calculating retention indices should compare with results from the literature in order to be able to use retention indices from the literature to confirm its own measurement results. The choice and optimization of the sample preparation method is also part of the method validation of drug screening. This procedure has already been described in Chapter 7.

8.2.1.7 Quality Assurance

Method validation is a procedure for checking the capabilities and possible weak points of a method, although the quality of the analysis as a routine operation must also be established. The detailed procedure is described in Chapter 13. As well as internal quality assurance, participation in interlaboratory testing is also of great importance as a method of external quality assurance for drug screening. With the aid of such interlaboratory tests, not only is the capability of the analytical equipment tested using substances in low doses or of rare occurrence, but the analyst's expertise is kept up to date on a regular basis. The results of an interlaboratory test also give a good overview of the effectiveness and application potential of the various methods of drug screening.

8.2.1.8 Mass Spectra

The mass spectra described here were obtained by electron impact ionization at a potential difference of 70 V, so that the energy of the electrons was 70 eV. If electrons with this energy collide with organic molecules, the extent of fragmentation is usually high and the reproducibility of the mass spectra is good [40]. Most drugs of abuse when subjected to electron impact ionization give very specific mass spectra, many of which contain key fragments from which a rapid identification of substances is possible (see Fig. 8-5).

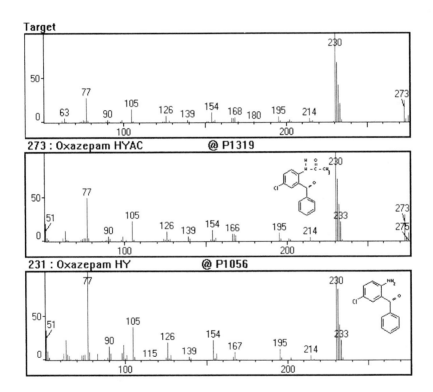

Fig. 8-5. Characteristic mass spectrum with the result of the library search

Based on its library search [41], the computer suggests the presence of the following substances in the unknown mass spectrum:

Hit	Significance	Molecular mass	Name
1	93	273	Oxazepam, hydrolyzed, acetylated
2	80	231	Oxazepam hydrolyzed
3	60	259	BDMPEA
4	55	272	Chlorphenoxamine-M, hydrolyzed, acetylated
5	54	315	Phenothiazine-M, twice acetylated

The good agreement with the proposal in first place enables a relatively certain identification of the unknown spectrum to be made. On the other hand, some mass spectra contain as their basis peak a less characteristic mass fragment whose size much exceeds that of the other mass peaks. In these cases, in spite of subtraction of the peak background, a false interpretation of the mass spectrum is possible, especially if compounds with appreciably different structures lead to these uncharacteristic spectra following similar fragmentation behavior (see Fig. 8-6). The results of the automatic library search should not therefore be used without careful thought. Their credibility should be checked, for example, by referring to the retention indices.

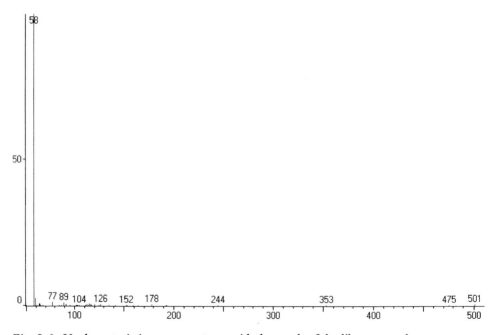

Fig. 8-6. Uncharacteristic mass spectrum with the result of the library search

Hit	Significance	Molecular mass	Name
1	93	337	Doxepin-M (hydroxy-) AC isomer -1
2	93	337	Doxepin-M (hydroxy-) AC isomer-2
3	92	58	Acetone
4	92	291	Tramadol-M (O-desmethyl-) AC
5	91	291	Melitracene
6	91	241	Benzalkonium chloride
7	91	339	Propoxyphene
8	90	213	Benzalkonium chloride
9	90	353	Dosulepin-M (hydroxy-) AC isomer-2
10	88	323	Cyamemazine

8.2.1.9 Evaluation

In drug screening, mass spectra which show intense peaks at m/z 58 or 72 and in which the remaining peaks have less than 3% of this intensity must be regarded as poorly characterized. As shown in Fig. 8-7, after a computerized library search the computer suggests structurally different substances with high significance rates that differ only slightly from each other.

In order to avoid coming to false conclusions in such cases, the following procedures are possible:

- Reference to the retention indices of the unknown substance
- A search for other peaks (metabolites) in the chromatogram which support the result
- Derivatization and remeasurement of the sample.

If the library search is limited to only a small number of suggestions, it is even possible for the computer not to include in its list of suggestions the substance actually present in the sample. Therefore, ten suggestions should be listed after any library search by the computer. In the case illustrated in Fig. 8-7, identification of the peak would probably be more reliable if a sample was derivatized with acetic anhydride (see Section 8.2.1.11) and injected again, as the acetylation product shows more characteristic masses than the starting material. Figure 8-8 shows the mass spectrum of the same peak after acetylation in comparison with the result of the library search. The substance suggested in first place (phentermine) differs very considerably from methamphetamine, which is suggested with high significance in Fig. 8-7.

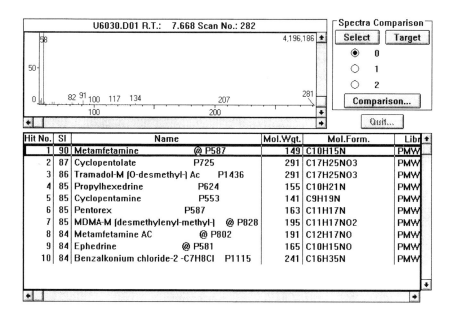

Fig. 8-7. Uncharacteristic mass spectrum with the result of the library search

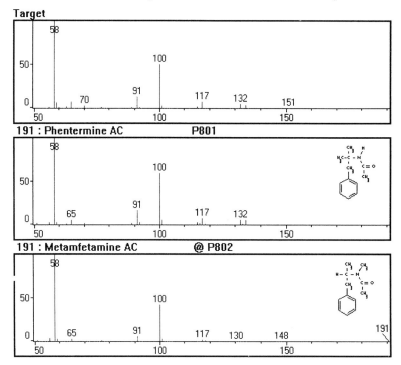

Fig. 8-8. Mass spectrum after acetylation with the result of the library search

Hit	Significance	Molecular mass	Name
1	94	191	Phentermine AC
2	92	191	Methamphetamine AC
3	90	249	Ephedrine 2AC
4	90	249	Pseudoephedrine 2AC
5	88	205	Pentorex AC
6	86	149	Methamphetamine
7	85	249	Pholedrine 2AC
8	83	307	Oxilofrine 3AC
9	82	291	Cyclopentolate
10	80	279	Oxilofrine ME 2AC

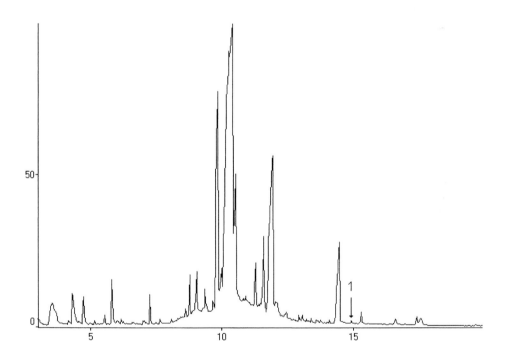

Fig. 8-9. Chromatogram of a sample with some highly concentrated substances:
1 = fentanyl

After injection of the prepared urine sample, saturation of the detector can occur rapidly with some peaks. The resulting mass spectra are less suitable for a library search, as the ratios of the individual masses to each other differ from those in the library, in which the spectra are measured at lower concentrations, and false results of the automatic library search can be obtained. Poorly separated peaks can also be a result of substance concentrations that are too high. A practical example is illustrated in Fig. 8-9. This urine sample was obtained from a patient who had been operated on a few hours earlier and who had received 250 µg fentanyl, as well as other substances, during the operation.

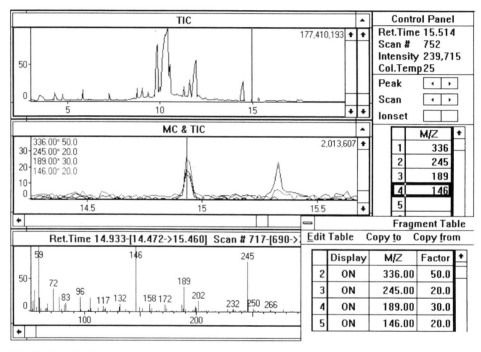

Fig. 8-10. Hard copy of a computer display for determining fentanyl with mass fragments

The best way to avoid mistakes in the identification of poorly separated peaks is to inject a sample of lower concentration. Alternatively, the retention times may also be taken into consideration to confirm the identification. These problems especially affect the evaluation of chromatograms of samples from polytoxicomanes who consume high doses, as in these cases both very small and very large peaks of relevant substances must be identified (e.g. benzodiazepines with opiates). Bearing in mind the wide variations in the nature of the abused substances or groups of substances and in the concentrations of these present in urine, the evaluation of the chromatograms should be performed manually.

One of the advantages of mass spectrometry over other chromatographic detection methods is that for the chromatogram evaluation significant mass fragments of individual substances can be represented in amplified form, enabling even superimposed signals to be identified. It is advantageous when dealing with this problem to prepare tables of mass results. The chromatogram shown in Fig. 8-9 was subsequently processed by the computer for the determination of fentanyl (see Fig. 8-10). This enables the identification of the relevant peaks to be considerably speeded up, as the number of peaks to be assessed can be minimized because the mass fragments have been preselected. Thus, the more specific the problem, the more effectively can the evaluation be automated. The characteristic mass fragments of the relevant substances are listed in Section 8.3.

This mass spectrum of fentanyl does not agree very well with the library spectrum, as several additional substances together with their metabolites are present in high concentrations in the sample. In this case, the sample can be measured again after optimization of the scanning range to suit the problem (m/z 130–345 Dalton). The result is shown in Fig. 8-11. The mass spectrum can now be identified with certainty.

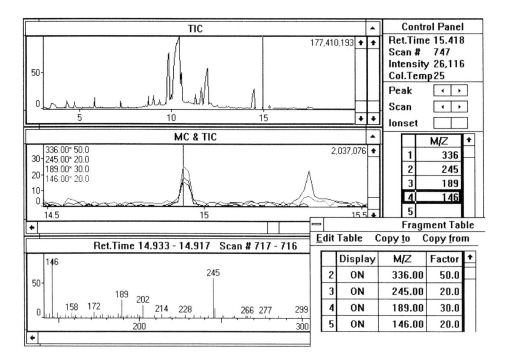

Fig. 8-11. Chromatogram with amplified mass fragments and the mass spectrum of fentanyl of the sample shown in Fig. 8-9 after optimization of the scanning range

The software supplied with commercial equipment often also supports the documentation of the sample evaluation. Identified peaks of a chromatogram can be stored in a table, and if a chromatogram needs to be looked at later, these data can be retrieved quickly. Because of the high measurement sensitivity of mass spectrometric detectors, carry-over of highly concentrated components such as dihydrocodeine can falsify the results from subsequent samples. To prevent this kind of unpleasant surprise, chromatograms should at least be evaluated in the chronological order of the samples. "Blank runs" or methanol injections after analyzing samples of this type minimize the risk of false positive results.

8.2.1.10 Sample Preparation, Glucuronide Breakdown

For all the chromatograms in this book from this point onwards, 1.5 ml urine was incubated with 20 µl β-glucuronidase at 56°C for 45 min and prepared by the solid phase extraction method described in Chapter 7. The partially evaporated eluates were taken up in 100 µl methanol and placed in a 1.5 ml PE vial, further evaporated to dryness, derivatized in accordance with the analysis required, and taken up again in 25 µl methanol. Assuming a one hundred per cent recovery rate, the concentration of the drug in the methanolic extract is 75 times as high as that in the starting material. Using this method, good recovery rates, i.e. between 80 and 100%, of basic substances are obtained. Neutral substances such as caffeine are partially eliminated by the methanol washing stage and therefore appear to a lesser extent (<50%) in the eluate. Liquid-liquid extraction (LLE) of neutral substances from urine, a process which also extracts acidic substances such as the barbiturates, gives considerably better recovery rates. However, it is of some importance that, because many neutral hormones and fatty acids are co-extracted from the urine, the detection of low-dose dependence, especially that of benzodiazepines, can be a problem, as the ranges of retention times of relevant benzodiazepine metabolites interfere with those of the physiologically neutral components. Only a targeted search for the characteristic mass fragments, e.g. by having an idea of what to look for, can solve the problem. More detailed information on sample preparation and derivatization can be found in Chapter 7.

8.2.1.11 Derivatization

Of the large number of different derivatization methods for the determination of medicaments and drugs, acetylation with acetic anhydride is one of the commonest. Reaction with acetic anhydride leads to volatile derivatives that reach the detector in higher yields than do the starting substances. Oxazepam, morphine and the polar carboxy-, hydroxy- or amino-metabolites of many medicaments are examples of this, so that the derivatization of such substances considerably increases their detection sensitivity, and although it may not be necessary to derivatize some substances, it can considerably improve the certainty of the identification of the mass spectrum, as explained in Section 8.2.1.9. For the broad spectrum of drug screening, acetylation is advantageous as a routine procedure, as the mass spectra of most derivatives of drugs of abuse are listed in the libraries of reference spectra. The reaction with pentafluoropropionic anhydride is another widely used derivatization method (see also Chapter 7.5), although the library lists of mass spectra of reaction products, especially those of the benzodiazepiness, are often less than comprehensive and need to be supplemented. The use of this derivatization method assumes that the mass range scanned can be extended to at least 600 Dalton (morphine-2PFP: 577 Dalton). The software for the GC/MS usually enables a laboratory-produced library of mass spectra to be used. In the descriptions of the mass spectra, the retention time of the peak and the method used for the measurement are also included. Thus it is not necessary to confirm a result by calculating the retention time from the retention index. The reaction with acetic anhydride causes the chromatogram of the measured sample to contain a large number of other peaks due to impurities (and sometimes also to the derivatization process) in addition to those of the desired derivatives. The temperature-pressure program of the GC/MS used must take account of this. If the acetic anhydride is mixed with pyridine as a proton absorber the number of interfering peaks very much increases. For this reason, acetic anhydride only is used in the process of derivatization and somewhat lower yields are tolerated.

This subject is discussed in more detail in the Sections on the individual groups of drugs, especially where important members of a group or their metabolites do not give satisfactory results using the method described here.

8.2.2 Screening Method A, without Derivatization

The choice of the GC/MS method depends on the precise nature of the analytical problem as well as on the method of sample preparation. Very many substances and their metabolites can be determined even in subtherapeutic concentrations without problems and without any derivatization on the basis of the sample preparation methods described above. This applies to the habit-forming drugs, e.g. chlormethiazole, the opiates/opioids codeine, dehydrocodeine, methadone, tilidine and tramadol, and cocaine and its metabolite methylecgonine. Substances such as amphetamine, ephedrine, morphine and the cocaine metabolite benzoylecgonine can usually be determined after high-dose consumption without derivatization by GC/MS, even on the following day. However, for lower urine concentrations, derivatization of the substances is essential in order to at least reach the limits of determination of immunological methods. The equipment and the chromatographic conditions used for drug analysis in our laboratory are listed in Tables 8-4 and 8-5.

Table 8-4. Chromatographic equipment for drug analysis

Gas chromatograph:	GC-17 (Shimadzu)
Mass spectrometer:	QP 5000
Software:	CLASS 5000
Libraries:	PMW-TOX 2, NIST 75000 and libraries produced by the user

Table 8-5. Chromatographic conditions for the measurement of underivatized urines (method A)

Injection volume:	1 µl, splitless		
Injector temperature:	260°C		
Oven temperature:	60°C		
Holding time:	1.5 min		
Temperature program:	Heating rate	End temperature	Holding time
	25°C/min	160°C	
	10°C/min	200°C	
	15°C/min	280°C	5.17 min
Carrier gas:	Helium		
Carrier gas pressure:	Pressure 30 kPa		
Linear velocity of carrier gas:	32.4 cm/s		
Pressure program:	Rate of increase	End pressure	Holding time
	16 kPa/min	93 kPa	
	7.2 kPa/min	125 kPa	
	9.5 kPa/min	170 kPa	5.38 min
Column:	SE-54 (CS Chromatographic Service), 25 m, 0.25 mm I.D., 0.25 µm film		
Interface temperature:	280°C		
Detector voltage:	2.0 kV		
Time for solvent removal:	2.5 min		

The advantage of this method lies in the saving of time. If the practical situation permits it, the derivatization step in the sample preparation can be omitted. The chromatogram of this underivatized urine as a rule contains up to 30% fewer peaks than the same urine treated with acetic anhydride, so that the analysis time can be reduced by 10 min compared with that for the derivatized urine. The smaller number of peaks also enables the time needed for the evaluation of the chromatogram to be reduced. The disadvantage that some important substances such as oxazepam and morphine cannot be determined with sufficient sensitivity without derivatization means that it would be extremely inadvisable to omit this step when carrying out drug screening (see Section 8.2.1.5).

In our laboratory, we also prefer to omit derivatization when confirming the results of analysis by HPLC/DAD. For the the latter method of analysis, solid phase extraction as described in Chapter 7 is carried out and the residue is taken up in 100 µl methanol. Of the remainder of the methanol extract for HPLC analysis, 1 µl is injected into the GC/MS.

8.2.3   Screening Method B, including Derivatization with Acetic Anhydride

Derivatization of the urine extract with acetic anhydride enables almost all the addictive substances recovered by means of the sample preparation process to be determined, at least down to the level of 100 ng/ml initially present in the urine. The only exception is the cocaine metabolite benzoylecgonine, as this can only be determined at this sensitivity by methylation or reaction with pentafluoropropionic anhydride. However, other cocaine metabolites, principally methylecgonine, are determined without difficulty by this method, so that the poor sensitivity of the benzoylecgonine method for determining cocaine consumption by urine analysis is not of any great significance.

As already discussed in Section 8.2.2, the number of peaks in a chromatogram increases considerably on derivatization. Figure 8-12 shows the chromatogram of a sample before and after acetylation. Method B (Table 8-6) will give increased separation using a slowly rising temperature-pressure program.

Fig. 8-12. Chromatogram of sample from a patient after consumption of tilidine,
before (above) and after (below) acetylation of the sample:
Above: 1 = Tilidine-M (TM) (phenylcyclohexanone); 2 = TM-bis-nor;
3 = TM-nor
Below: 1 = Tilidine-M (TM) (phenylcyclohexanone); 2 = TM-artifact AC;
3 = TM-bis-nor AC; 4 = TM-bis-nor-oxime; 5 = TM-nor AC;
6 = TM-bis-nor-hydroxy

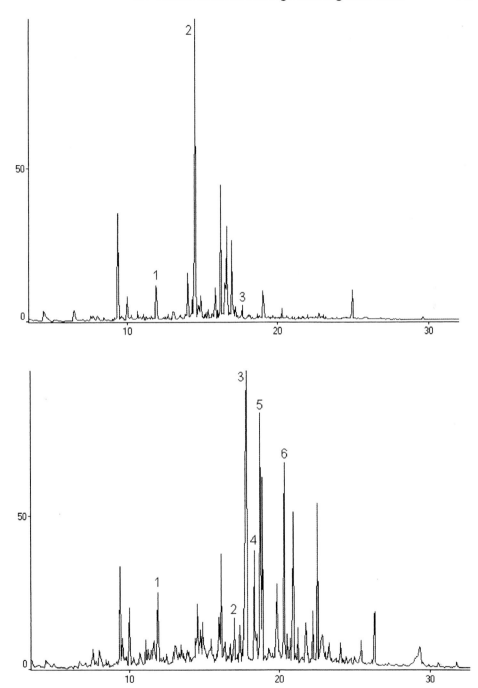

Fig. 8-12. (For legend see p. 96)

Table 8-6. Chromatographic conditions for the measurement of acetylated urines
          (Method B)

Injection volume:	1 µl, splitless
Injector temperature:	280°C
Oven temperature:	70°C
Holding time:	1.5 min

Temperature program:	Heating rate	End temperature	Holding time
	20°C/min	160°C	
	5°C/min	250°C	
	20°C/min	300°C	6.0 min

Carrier gas:	Helium
Carrier gas pressure:	40 kPa
Linear velocity of carrier gas:	27.5 cm/s

Pressure program:	Rate of increase	End pressure	Holding time
	10 kPa/min	90 kPa	
	3 kPa/min	140 kPa	
	13 kPa/min	180 kPa	6.26 min

Column:	SE-54 (CS Chromatographic Service), 25 m, 0.25 mm I.D., 0.25 µm film
Interface temperature:	300°C
Detector voltage:	2.0 kV
Time for solvent removal:	2.5 min
Duration:	3–32.5 min
Masses measured:	55–509 Dalton
Scanning interval:	0.92 s

8.3 Determination of Addictive Substances and Groups of Substances

Possible methods for analyzing addictive substances and the usual accompanying medication are described below. Apart from Δ^9-THC, only basic substances or groups of substances are discussed. All substances except the Δ^9-THC metabolites can be extracted by the sample preparation method described in Chapter 7. Unless otherwise stated, in the GC/MS chromatograms shown, 1.5 ml urine was prepared by the solid phase extraction process described, the residue was taken up in 20 µl methanol, 1 µl of this was injected without derivatization, and this was measured by method A (see Table 8-5). In the various Sections, chromatographic methods of determination are compared with the immunological alternatives. This comparison is very important, as the chromatographic confirmation of immunological drug screening results, which is a widely used practical technique, requires detailed information about the immunological method used. Detailed knowledge of the threshold concentrations of the relevant substances is therefore very important.

The list of the immunological reagents given is not claimed to be in any way complete. We emphasize that comparable results can sometimes be obtained by other immunological methods not described here. Having regard to the large number of rapid immunological tests, we have, in the interests of clarity of the comparative tables, selected one product of this type to compare with the other methods. Information about the cross-reactivities of the various substances towards the immunological tests has been obtained from the current information sheets provided by the manufacturers.

8.3.1 Opiates/Opioids

Opioids are those substances that can activate localized opioid receptors in the pain-controlling system of the body. Their action can therefore be antagonized by naloxone. From a clinical point of view, morphine is the most important substance of this group. It is obtained from opium, which is the brownish latex obtained from cuts made in the unripe but fully grown fruits of *Papaver somniferum*, the opium poppy. Opium contains not only morphine but also over thirty other alkaloids (Fig. 8-13) including the isoquinoline alkaloids codeine, papaverine, noscapine, thebaine and others, which make up ca. 25% of the opium. The alkaloids in opium are sometimes present as salts of meconic acid. The opiates form a separate subgroup within the opioids whose member

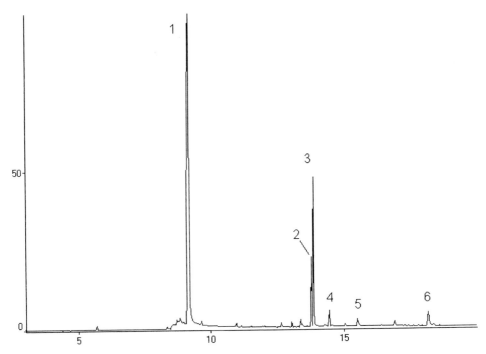

Fig. 8-13. Chromatogram of a heroin preparation dissolved in methanol.
        1 = caffeine; 2 = codeine-AC; 3 = monoacetymorphine;
        4 = diacetylmorphine; 5 = papaverine; 6 = noscapine

substances can be directly structurally derived from morphine (heroin, codeine, dihydrocodeine etc.).

The synthetic opioids, whose basic structure shows a very important similarity to morphine, i.e. a phenylpropylamine configuration [42] with an asymmetric carbon atom in close proximity to aromatics, are pharmacodynamically different only in respect of the central and peripheral effects in their characteristic spectrum. Apart from the opiates mentioned, only tramadol and tilidine are abused in significant quantities. All the opioids are of therapeutic value, primarily because of their analgesic and antitussive effects. The usual daily dose rates of the substances discussed here, depending on the strength of their effects, lie between 20 mg (morphine) and 400 mg (tramadol), but can be greatly increased, especially in pain control therapy of malignant diseases. Products derived from pethidine and fenatyl are described in Section 8.3.4.

8.3.1.1 Metabolism

The opioids mentioned above are mainly excreted renally (via the kidneys). The proportion excreted in the bile (via the gall bladder) always lies approximately between 10 and 20%. Figure 8-14 shows the important steps in the metabolism of some opiates.

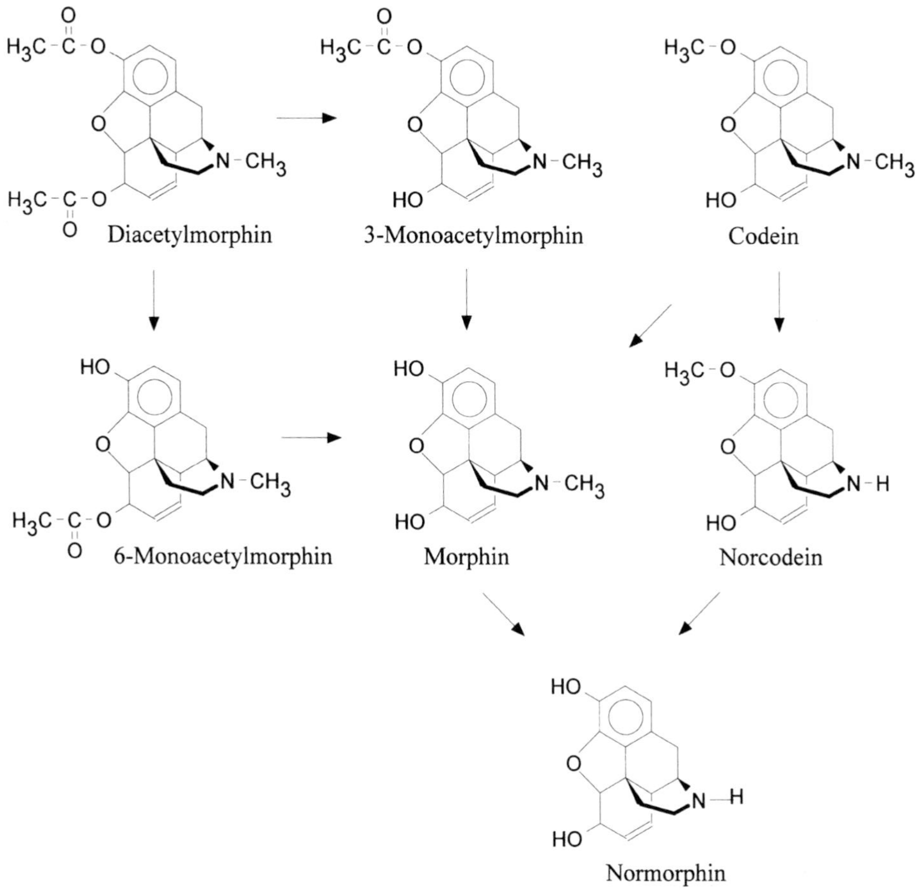

Fig. 8-14. Some important stages in the metabolism of opiates (without glucuronides)

- *Codeine* is excreted via the kidneys mainly unchanged or as the conjugate with glucuronic acid. Amounts of norcodeine (10-20%) and morphine (5-15%) considerably exceed the amounts of the other metabolites (normorphine, hydrocodone, dihydrocodeine), which are present only in traces [43, 44].

- *Dihydrocodeine (DHC)* is more extensively used than codeine as a substitution drug by heroin addicts. Prescribed weekly rations can sometimes be as high as 2000 ml of a 2.5% solution, i.e., 50 g per week. The concentrations found in the urine are correspondingly high, and this rapidly leads to saturation of the detector. So far, this substance is not permitted for drug substitution, and control tests to rule out possible co-consumption are not required legally. The metabolism of DHC proceeds in analogy to that of codeine via O- and N-demethylation to nordihydrocodeine and dihydromorphine. In contrast to codeine, which is almost completely excreted in the urine within 24 h, the elimination of DHC is slower, and intoxication with this substance is therefore more difficult to treat than heroin intoxication [45].
- Elimination of *heroin* (diacetylmorphine) is 80% renal. Metabolic inactivation mainly gives morphine-3-glucuronide together with a few percent of the metabolites 6-monoacetylmorphine, non-conjugated morphine and normorphine. Unchanged diacetylmorphine can be detected only in trace amounts if at all. The hydrolysis of heroin via 3-acetylmorphine takes place to a small extent only [44].
- After the consumption of *morphine*, it is rapidly resorbed and distributed in the organism. Up to 90% of the dose is excreted renally within the first 24 h, mainly as glucuronides, and 10% is eliminated via the bile in the feces. Normorphine seldom comprises more than 1% of the amount of substance present in the urine. The percentage of free morphine present increases in parallel with the acidity of the urine [46].
- *Noscapine and papaverine*, like morphine, are members of the group of isoquinoline alkaloids, but are not opioids. However, both of these substances are discussed here as they are important constituents of opium and occur when this is processed to obtain heroin. Percentages of these two alkaloids in heroin vary greatly, in many cases enabling it to be classified in various grades. Therapeutically, noscapine is used as an antitussive, but papaverine is seldom used for this purpose because of its spasmolytic and vasodilating effects. After heroin is taken, considerable quantities of these two substances sometimes appear in the organism. The metabolites, formed by demethylation and conjugation with glucuronic acid and sulfate, are mainly excreted renally.
- *Tilidine*, after being taken orally, undergoes an intensive first-pass effect to form nortilidine, the actual active substance. Other metabolites include bisnortilidine and hydroxylation and conjugation products (see Fig. 8-12). Of the amount taken, 10% is excreted in the bile. The commercial preparation, for oral application, also contains a proportion of *naloxone* to prevent abuse

by intravenous injection. The *N*-desalkyl- and 6-hydroxyderivatives of this substance are excreted renally [46].

- *Tramadol*, after being taken orally, is almost completely resorbed. Metabolism takes place by demethylation and conjugation with glucuronic and sulfuric acids. *O*-desmethyltramadol has a stronger analgesic effect than that of tramadol, but all other *O*- and *N*-desmethyl metabolites are pharmacologically inactive. In humans, the proportion of tramadol excreted unchanged renally is ca. 30% of the applied dose [46].

8.3.1.2 Analysis

The determination of opioids is one of the most important tasks of drug screening. In substitution therapy with methadone, it is necessary to ensure that these substances are not being used at the same time as the methadone. All the immunological reagents are able to detect the opiates, but only as a group of substances, and the task of distinguishing between the individual substances or detecting the other opioids must be by means of chromatography. This is especially important when heroin addicts are given codeine or DHC, a widely used substitution therapy, as any co-consumption of opiates in addition to these heroin substitutes cannot be detected analytically by the use of immunological screening methods. The cross-reactivities of various immunological methods for determining individual opioids are listed in Table 8-7, and, although the sensitivities, after glucuronide cleavage, are always high enough to detect the most important opiates in urine, positive measurement results obtained using immunology should only be regarded as "very probably positive", as false positive results which cannot be confirmed by chromatography occur again and again [47].

Table 8-7. Cut-off values of some opioids, substances naturally associated with them and their metabolites found in urine [µg/ml]

Substance	CEDIA®DAU	Emit®d.a.u.	dau-TRAK®	Triage™ 8®
Codeine	0.24[a]	1.0	0.38	0.3
Dihydrocodeine	0.6	0.26	0.4	0.3
6-Monoacetylmorphine	0.37	0.38	0.1	–
Morphine	0.3	0.3	0.3	0.3
Morphine-3-glucuronide	0.37	3.0	0.3	0.49

Table 8-7 (Cont.)

Substance	CEDIA®DAU	Emit®d.a.u.	dau-TRAK®	Triage™ 8®
Norcodeine	–	–	–	15
Normorphine	–	–	35	25
Noscapine	–	1000	1000	>100
Papaverine	–	1000	1000	>100
Tilidine	–	–	–	–
Tramadol	–	–	–	–

[a] The figures for the cross-reactivities are taken from the information provided by the manufacturers. Cross-reactivities not provided in this information are indicated by a dash (–). This symbol has the same meaning in subsequent Tables.

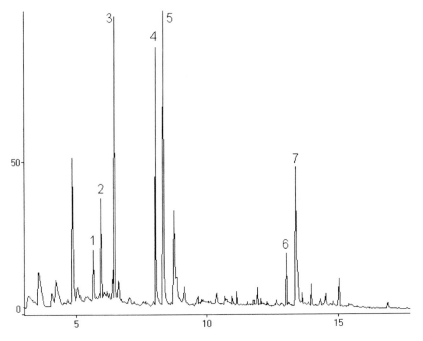

Fig. 8-15. Chromatogram of the urine sample of a polytoxicomane.
1 = nicotine; 2 = cocaine-met. artifact; 3 = methylecgonine; 4 = cotinine;
5 = meconine; 6 = codeine; 7 = morphine

After cleavage of conjugates, all substances and their metabolites in the chromatogram can be detected by GC/MS. Morphine can be determined without derivatization because of its two free hydroxyl groups, but only after high doses of morphine or heroin have been consumed and if concentrated pure substances are used in the analysis (see Fig. 8-15). The limits of detection of the other relevant opioids (50–100 ng/ml urine) are considerably better than those achieved by immunology even without derivatization, but it is strongly recommended that derivatization of morphine, e.g. with acetic anhydride, should be performed to give sensitive enough detection. This technique also increases the sensitivity towards DHC and codeine.

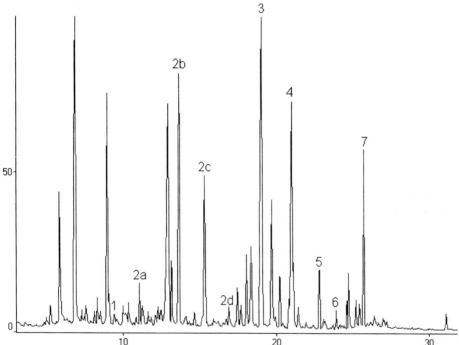

Fig. 8-16. Chromatogram of a urine after derivatization with acetic anhydride.
1 = meconine; 2a-d = tramadol-met. acetylated; 3 = codeine; 4 = acetyl codeine; 5 = diacetylmorphine; 6 = temazepam acetylated; 7 = norcodeine acetylated

After consumption of heroin, meconine, papaverine and noscapine can be detected in the sample material by GC/MS. This information is also useful for differentiating between illegal heroin consumption and the possible therapeutic consumption of a morphine-based proprietary medicine. Table 8-8 below shows

the characteristic mass fragments of the important underivatized opioids and
their metabolites and associated compounds.

Table 8-8. Opiates, their accompanying substances and metabolites without
derivatization

RI	Name	Characteristic masses with intensities		
2375	Codeine	299_{100}	229_{26}	
2410	Dihydrocodeine	301_{100}	244_{10}	
2400	Dihydrocodeine-M (O-desmethyl-)	287_{100}	230_{14}	
1780	Meconine	194_{92}	176_{52}	165_{100}
2455	Morphine	285_{100}	268_{16}	
2500	Heroin-M (3-acetyl-morphine)	327_{100}	310_{10}	285_{73}
2535	Heroin-M (6-acetyl-morphine)	327_{100}	268_{53}	124_{13}
3130	Noscapine	412_{1}	220_{100}	
2820	Papaverine	339_{83}	338_{100}	324_{84}
2805	Papaverine-M (O-desmethyl-)	325_{77}	324_{100}	310_{80}
1835	Tilidine	273_{1}	176_{7}	97_{100}
1820	Tilidine-M (nor-)	259_{7}	83_{100}	
1840	Tilidine-M (bis-nor-)	245_{6}	83_{29}	69_{100}
1950	Tilidine-M (bis-nor-hydroxy-)	261_{15}	85_{100}	
1945	Tramadol	263_{12}	58_{100}	
1975	Tramadol-M (O-desmethyl-) $-H_2O$	231_{3}	184_{6}	58_{100}
2200	Tramadol-M (hydroxy-)	279_{8}	234_{2}	58_{100}

The chromatogram in Fig. 8-16 to some extent reflects the multiplicity of
relevant peaks which occur when drug screening using the derivatized urine of
polytoxicomanes.

When the urine extract is derivatized with acetic anhydride, morphine is
converted into the two monoacetyl derivatives and diacetylmorphine, and
codeine and DHC are converted into the analogous 6-acetyl esters (Table 8-9).
The metabolites of noscapine and papaverine, like those of morphine, can be
determined with adequate sensitivity only after derivatization.

Reaction with pentafluoropropionic anhydride as the derivatization reagent
gives compounds with high and very characteristic masses (Table 8-10). When
this method of analysis is used, the mass range has an upper limit of at least

600 Dalton, while the determination of acetylated opioids only requires the scanning of masses up to 500 Dalton.

Table 8-9. Acetylation products of the opioids and opium alkaloids detectable in urine

RI	Name	Characteristic masses with intensities			
2500	Codeine-6-AC	341_{100}	282_{40}		
2620	Diacetylmorphine (heroin)	369_{59}	327_{100}	310_{36}	
2490	Heroin-M (3-acetylmorphine)	327_{100}	310_{10}	285_{73}	215_{19}
2535	Heroin-M (6-acetylmorphine)	327_{100}	268_{53}	162_{11}	124_{13}
2435	Dihydrocodeine-6-AC	343_{100}	300_{33}		
2490	Dihydrocodeine-M (O-desmethyl-) AC	329_{100}	287_{56}		
2700	Dihydrocodeine-M (nor-) AC	329_{40}	87_{10}		
2910	Papaverine-M (O-desmethyl-) AC	367_{47}	324_{100}	308_{26}	
2100	Tilidine-M (bis-nor-) AC	287_{53}	111_{89}	69_{100}	
2165	Tilidine M (nor-) AC	301_{3}	125_{100}	83_{74}	
2295	Tramadol-M (N-desmethyl-) –H$_2$O AC	273_{30}	200_{100}	86_{98}	
2465	Tramadol-M (bis-desmethyl-) –H$_2$O AC	301_{28}	228_{17}	86_{100}	

Table 8-10. Pentafluoropropionylation products of the opioids and opium alkaloids detectable in urine

RI	Name	Characteristic masses with intensities		
2430	Codeine PFP	445_{100}	282_{73}	
2360	Codeine-M (o-desmethyl-) 2PFP	577_{51}	414_{100}	
2360	Dihydrocodeine PFP	447_{100}	390_{16}	
2490	Heroin-M (3-acetylmorphine) PFP	473_{25}	268_{100}	
2650	Heroin-M (6-acetylmorphine) PFP	473_{90}	430_{11}	414_{100}
2360	Morphine 2PFP	577_{51}	414_{100}	

The chromatogram in Fig. 8-17 shows some derivatization products with pentafluoropropionic anhydride (PFP) from the urine of a consumer of cocaine and heroin.

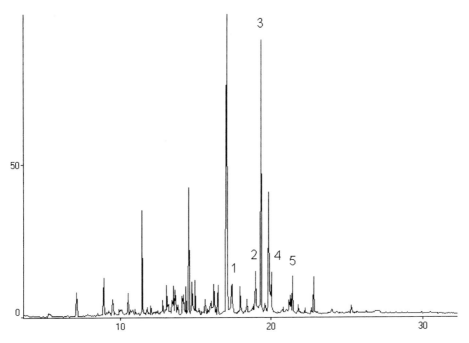

Fig. 8-17. Polytoxicomania: chromatogram of a urine sample after reaction with PFP. 1 = benzoylecgonine-PFP; 2 = cocaine; 3 = morphine-2 PFP; 4 = codeine-PFP; 5 = 3-acetylmorphine-PFP

The opioids and their metabolites can be identified very specifically in the acetonitrile gradients of HPLC/DAD. Even structurally similar compounds such as codeine and dihydrocodeine can be clearly differentiated, although, in analogy to GC/MS, it can be necessary here also to measure highly concentrated samples a second time after dilution. After the sample preparation, in contrast to GC/MS, no further steps are necessary. When heroin is consumed, not only do its metabolites appear in the HPLC chromatogram, but so also do the characteristic peak patterns of the variously metabolized substances noscapine and papaverine, whose presence in the urine increases the certainty of the analytical result. The sensitivity of detection is sufficient to resolve questions concerning results of immunological tests, so that this method can also be used as a universal method for the routine analysis of opioids.

8.3.2 Benzodiazepines

The benzodiazepines form a group of over 40 substances, approximately half of which are available in Germany in proprietary medicines. All benzodiazepines have anticonvulsant, anxiolytic, muscle-relaxant and sedative/hypnotic properties, although the various substances differ from each other with respect to the strengths of these effects. Most substances are available as tranquilizers/hypnotics and some as anticonvulsants or muscle relaxants. A habit-forming potential is common to all benzodiazepines, and a small number of these substances are abused extremely frequently. These are discussed further below.

As well as the opioids cocaine and cannabis, polytoxicomane drug abusers also consume bromazepam, diazepam and flunitrazepam in high doses, and fatalities among drug abusers and substitution patients can often be blamed on the consumption of these substances, often in combination with alcohol [48]. Patients who suffer low-dose dependence, in addition to consuming the three above-named substances, also consume therapeutic amounts of dipotassium clorazepate, flurazepam, lorezepam, nitrazepam or oxazepam, sometimes also in combination with amphetamines or antihistamines.

8.3.2.1 Metabolism

Benzodiazepines are metabolized in various ways, mainly involving reduction steps, dealkylation, hydroxylation in position 3, and conjugation with glucuronic acid. The glucuronides should be cleaved enzymatically during sample preparation to optimize the sensitivity of determination [49]. Table 8-11 lists elimination half-life times of the most important benzodiazepines and some of their metabolites.

The most important benzodiazepine is *diazepam*, which is one of the longest-acting benzodiazepines, having a plasma half-life of 20–100 h. On demethylation, desmethyldiazepam (= nordazepam) is formed, and this is then hydroxylated to form *oxazepam* (Fig. 8-18). A breakdown to oxazepam via temazepam also takes place to a small extent. All the metabolites are pharmacologically active and are present in the urine as glucuronides. Unchanged diazepam is present in the urine in trace amounts only. The long half-life of the first metabolites enables them to be detected in the urine for a long period, which can be up to 3 weeks after the last use of the drug at high dose rates. The extent of elimination via the bile is 10% of the consumed amount.

Bromazepam, in analogy to diazepam, is first broken down to the 3-hydroxy-derivative, which is the main metabolite in the urine. Later reactions, involving cleavage of the 1,4-benzodiazepine configuration, lead to 2-amino-3-hydroxy-5-bromobenzoyl pyridine and 2-(2-amino-5-bromobenzoyl) pyridine, both of which have a benzophenone structure. All the metabolites form conjugates with glucuronic acid. Bromazepam that has not been deactivated can sometimes be detected in the urines of polytoxicomane patients (see Fig. 8-19).

Table 8-11. Plasma half-life times of some important benzodiazepines and their
metabolites

Name	Half-life time [h]
Bromazepam	10–20
Diazepam	20–100
Desmethyldiazepam	40–100 (200)
Oxazepam	4–15 (25)
Flunitrazepam	10–20 (70)
7-Aminoflunitrazepam	20–30
Flurazepam	1–3
N-(2-hydroxyethyl)-flurazepam	10–20
Desalkylflurazepam	20–130 (280)
Lorazepam	12–16
Midazolam	2
α-Hydroxymidazolam	1
Nitrazepam	18–38

The metabolic deactivation of *flunitrazepam* illustrated in Fig. 8-20 proceeds via demethylation and hydroxylation to give desmethylflunitrazepam, an active metabolite, and 3-hydroxyflunitrazepam. However, these two compounds are only important in blood analysis. The main metabolite in the urine is the 7-amino-derivative, which is sometimes present as the acetamido-compound. 7-Aminodesmethylflunitrazepam and 3-hydroxy-7-acetamidoflunitrazepam are also relevant flunitrazepam metabolites in the urine, but the starting substance itself does not normally appear. Approximately 10% of the dose is excreted via the bile with the feces.

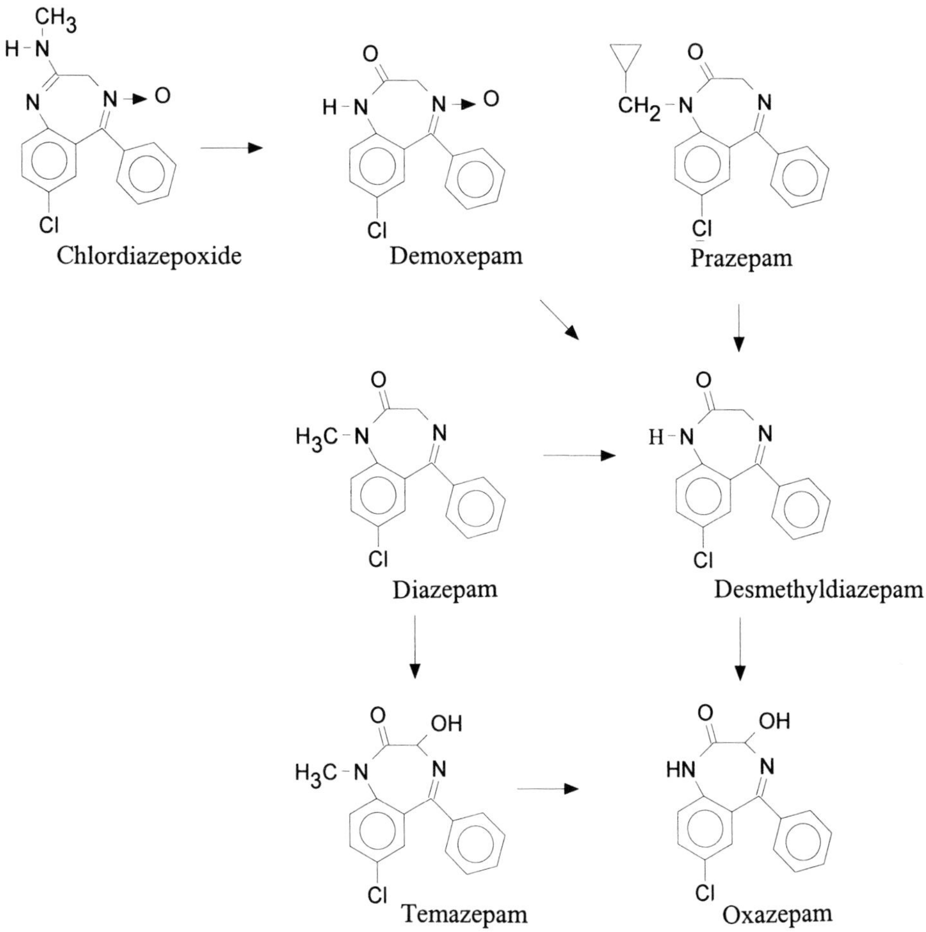

Fig. 8-18. Metabolism of some important benzodiazepines

The short half-life of *flurazepam* is due to a marked first-pass effect. The main metabolite in the urine is *N*-(2-hydroxyethyl)-flurazepam (Fig. 8-21), which, with desalkylflurazepam, are the pharmacologically active substances responsible for the cumulative effects of the starting substance. Didesethylflurazepam can also be detected in the urine.

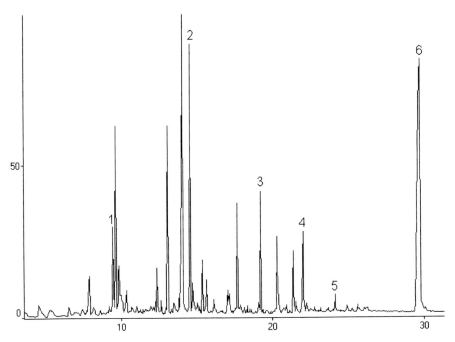

Fig. 8-19. Chromatogram of the sample from a patient with amphetamine and
benzodiazepine abuse.
1 = d-norpseudoephedrine; 2 = etophylline; 3 = bromazepam HY; 4 =
morphine; 5 = bromazepam; 6 = cafedrine

Other flurazepam derivatives not shown in this scheme (Fig. 8-21), e.g.
produced by oxidation of the alkyl side chain to the carboxylic acid derivative,
are quantitatively insignificant. All metabolites are to some extent present as
conjugates with glucuronic or sulfuric acid. The proportion of the dose excreted
in the bile, as is normal for most benzodiazepines, is ca. 10%.

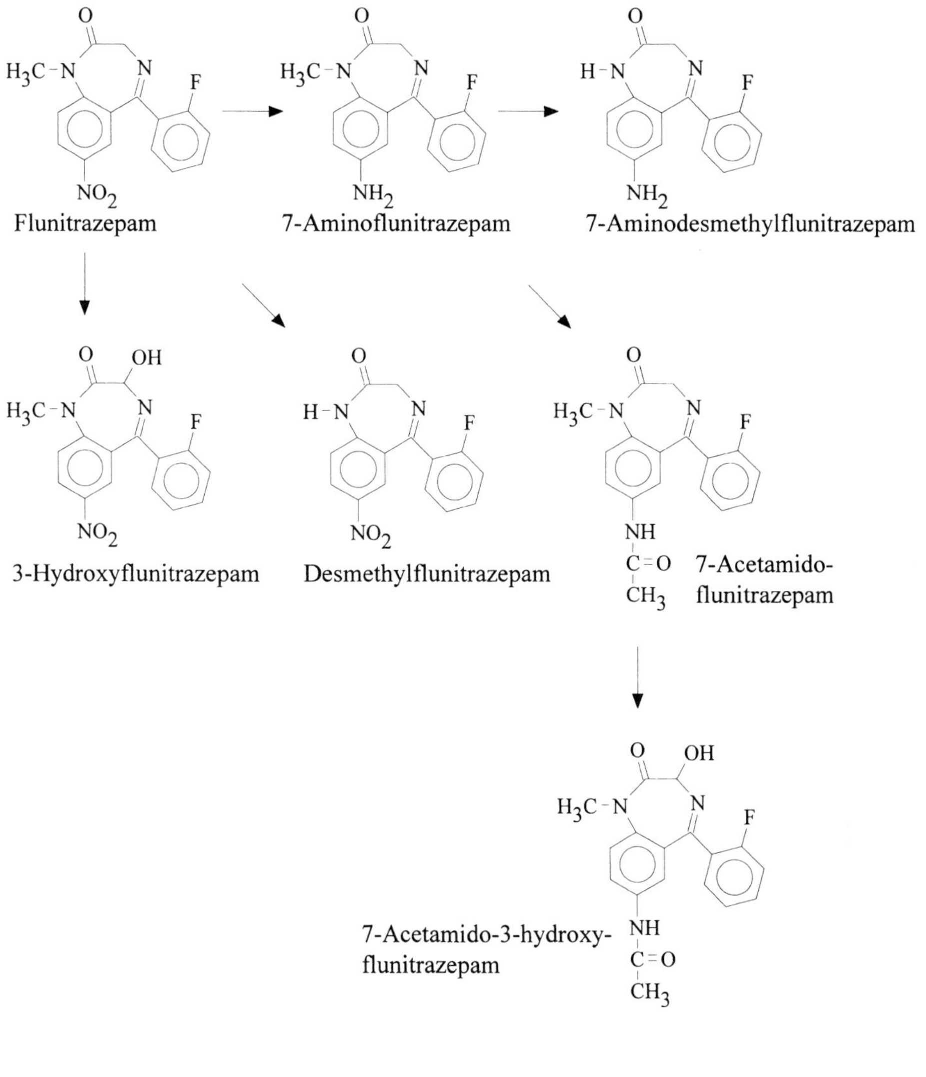

Fig. 8-20. Metabolism of flunitrazepam

Lorazepam is mainly excreted in the form of its glucuronide. As can be seen from Fig. 8-22, other metabolites such as quinazoline and hydroxyphenyl derivatives can be detected as well as small amounts of the unchanged substance [44]. It should be noted that all signals are very small compared with the other peaks.

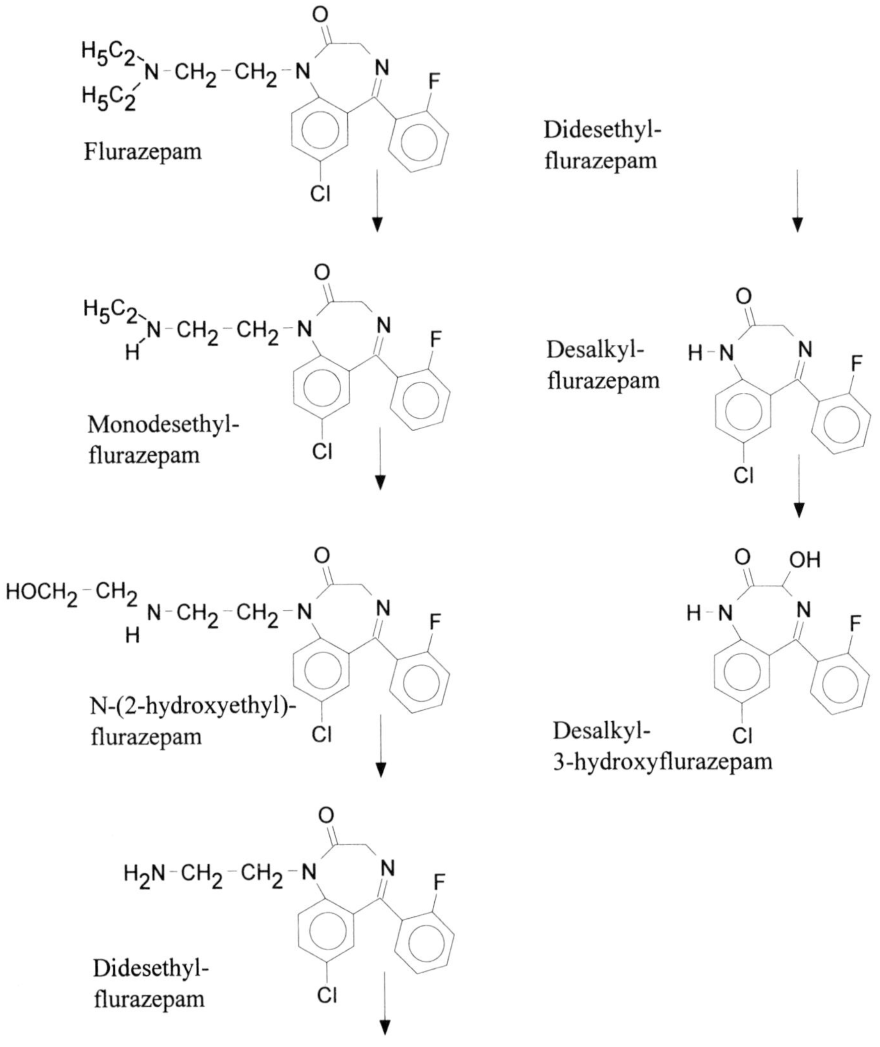

Fig. 8-21. Overview of the metabolism of flurazepam (part)

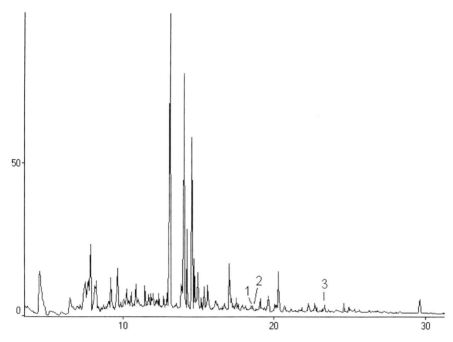

Fig. 8-22. Chromatogram of a urine after high-dose lorazepam consumption.
1 = lorazepam artifact; 2 = lorazepam HY; 3 = lorazepam

The main metabolites of *nitrazepam* are 7-aminonitrazepam, 7-acetamido-nitrazepam, and a benzophenone derivative. The proportion of glucuronides is comparatively small.

8.3.2.2 Analysis

The problems of benzodiazepine abuse have been known about for a long time. Nevertheless, the extent of the abuse of these substances and the consequences of their pharmacodynamic interactions with other drugs and medicaments are often underestimated. Consequently, the problems of determining these substances and their derivatives in urine accurately and with good sensitivity are not appreciated to an equal extent by all analysts or by all who submit samples to them.

Table 8-12. Cross-reactivitiesof some benzodiazepines and their metabolites in urine [μg/ml]

Substance	CEDIA®DAU	Emit®d.a.u.	dau-TRAK®	Triage™ 8®
Bromazepam	0.27	0.4	0.3	2.5
3-Hydroxybromazepam	–	–	–	–
2-Amino-3-hydroxy-5-bromobenzoylpyridine	–	–	–	–
2-(2-Amino-5-bromobenzoyl)pyridine	–	–	–	–
Diazepam	0.04	0.04	0.3	0.35
Desmethyldiazepam	0.07	0.06	0.2	1.0
Desmethyldiazepam glucuronide	–	–	–	0.3
Oxazepam	0.25	0.2	0.9	0.75
Oxazepam glucuronide	>1000	–	–	0.6
Flunitrazepam	0.14	0.1	0.15	0.3
Desmethylflunitrazepam	–	–	–	–
3-Hydroxyflunitrazepam	–	–	10	–
7-Aminoflunitrazepam	–	–	20	7.5
7-Acetamido-flunitrazepam	–	–	–	–
7-Amino-1-desmethyl-flunitrazepam	–	–	–	–
Flurazepam	0.08	0.1	20	0.75
Monodesethylflurazepam	–	–	6.0	–
N-(2-hydroxyethyl)-flurazepam	–	0.06	1.2	–
Didesethylflurazepam	–	–	3.0	–
Desethylflurazepam	0.06	0.1	0.7	1.5
Desalkyl-3-hydroxyflurazepam	–	–	–	–
Lorazepam	0.17	1.0	2.0	5.0
Lorazepam glucuronide	>1000	–	–	0.4
Midazolam	–	0.07	0.3	1.0
α-Hydroxymidazolam	–	–	–	–
Nitrazepam	0.2	0.2	40	0.5
7-Aminonitrazepam	–	–	2.2	–
7-Acetamidonitrazepam	–	–	–	–

In analogy to the opiates, the benzodiazepines have hitherto been determined by immunological methods as a substance group only. While the four most important opiates and their metabolites are always determined with high sensitivity, in the case of the most important benzodiazepines this is only true for those substances whose metabolism takes place via desmethyldiazepam and oxazepam (see Fig. 8-18) and for flurazepam, provided that sample preparation is always carried out. For example, although flunitrazepam can be determined by immunological methods, as indicated in Table 8-12, its main metabolite in urine, the 7-amino-derivative, is only likely to give a positive measurement signal in very high concentrations, i.e. 10–20 µg/ml. The same is true for bromazepam, which also can be detected immunologically only at high dose levels. Concrete figures for the detection sensitivity of the bromazepam metabolites cannot be obtained from the information sheets provided by the manufacturers, and the stuation is not helped by the fact that two metabolites of bromazepam, being benzophenone derivatives, no longer have the characteristic structure of the 1,4-benzodiazepines and therefore do not possess a suitable antigen structure for this method of analysis. Thus, the two highly relevant substances bromazepam and flunitrazepam, together with lorazepam, cannot be accurately determined in the range of low-dose dependence by the immunological method.

HPLC/DAD is very important in benzodiazepine analysis. After enzymatic cleavage of the conjugate, the individual metabolites of the benzodiazepines can be clearly distinguished from each other and identified in the chromatogram, so that, for example, the co-consumption of bromazepam when diazepam therapy is in progress can be detected. In sample material of low concentration, for example, the determination of several metabolites of the same starting substance considerably improves the quality of the analytical results. The limit of detection of HPLC is more than adequate in the region of 10–20 ng/ml for routine and confirmatory analysis.

The use of GC/MS in benzodiazepine drug screening usually necessitates a derivatization step during sample preparation if this method of analysis is to be satisfactory for low dose ranges. Benzodiazepines are usually excreted as polar hydroxy- or amino-compounds, which undergo strong interactions with the materials of the injector and the column film. Therefore, only small proportions of these substances reach the detector in underivatized form. A chromatogram typical of these substances is shown in Fig. 8-23. With high dose rates of diazepam in the case of a polytoxicomane, considerably larger amounts of desmethyldiazepam reach the detector than of oxazepam.

If a measurement of benzodiazepines in the low therapeutic dose rate area is required, acetylation is generally advisable. Table 8-13 lists some substances which can be determined after high dose rates without derivatization.

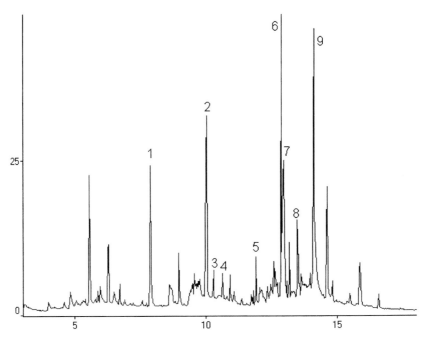

Fig. 8-23. Chromatogram of the urine of a diazepam user.
1 = cotinine; 2 = carbamazepine-M. artifact; 3 = oxazepam-HY;
4 = oxazepam-M. artifact; 5 = doxepin-M.; 6 = codeine; 7 = norcodeine;
8 = desmethyldiazepam; 9 = nalorphine

Table 8-13. Benzodiazepines and their metabolites that can be determined in urine without derivatization

RI	Name	Characteristic masses with intensities			
2670	Bromazepam	315_{91}	236_{100}		
2250	Bromazepam HY	276_{27}	247_{100}		
2470	Bromazepam-M (3-hydroxy-)	313_{100}	284_{22}		
2520	Diazepam-M (nor-)	270_{86}	269_{97}	242_{100}	241_{82}
2050	Diazepam-M (nor-) HY	231_{80}	230_{95}	77_{100}	
2400	Diazepam-M (nor-hydroxy-) HY	247_{72}	246_{100}		
2750	Diazepam-M (nor-hydroxy-)	286_{82}	258_{100}		
2615	Flunitrazepam-M (amino-)	283_{100}	255_{78}		
2470	Flunitrazepam-M (desalkyl-)	288_{100}	260_{93}	259_{57}	

Table 8-13 (Cont.)

RI	Name	Characteristic masses with intensities		
2650	Flunitrazepam-M (bi-desethyl-)–H$_2$O	313_{100}		
2440	Lorazepam	302_{45}	239_{91}	75_{100}
2140	Lorazepam artifact	288_{57}	287_{54}	253_{100}
2580	Midazolam	325_{47}	310_{100}	
2830	Midazolam-M (amino-)	341_{11}	310_{100}	
2785	Nitrazepam-M (amino-)	251_{100}	223_{66}	
2320	Oxazepam	268_{98}	77_{100}	
2050	Oxazepam HY	231_{80}	230_{95}	77_{100}
2070	Oxazepam artifact - 1	240_{59}	239_{100}	205_{81}
2625	Temazepam	300_{33}	271_{100}	
2050	Temazepam-M HY	231_{80}	230_{95}	77_{100}

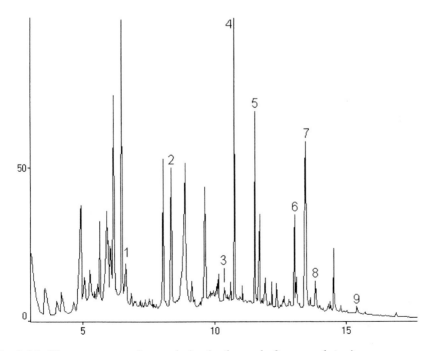

Fig. 8-24. Chromatogram of an underivatized sample from a polytoxicomane.
1 = methylecgonine; 2 = meconine; 3 = EMDP; 4 = EDDP; 5 = methadone;
6 = codeine; 7 = morphine; 8 = 7-aminoflunitrazepam; 9 = papaverine-M

Compared with the opioids, benzodiazepines are consumed in relatively low doses, the daily dose seldom exceeding 100 mg. Consequently, benzodiazepines and their metabolites usually give much smaller signals in the chromatogram than those of the opioids and are sometimes unrecognized. The chromatogram of the urine sample from a polytoxicomane patient (Fig. 8-24) illustrates this problem. The signals from the opioids etc. are far bigger than those from 7-amino-flunitrazepam.

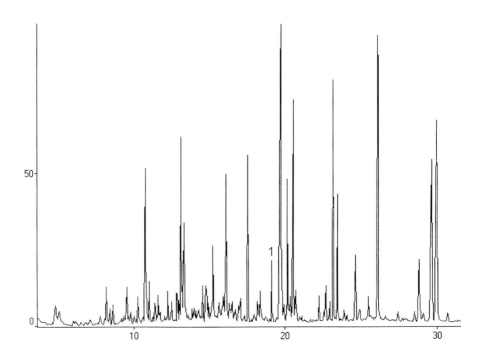

Fig. 8-25. Chromatogram of an oxazepam urine after acetylation.
          1 = oxazepam acetylated

Acetylation of the benzodiazepines leads to the analogous acetamido- and acetyl-derivatives, whose characteristic mass fragments are listed in Table 8-14. After acetylation, the limit of detection by GC/MS (around 30 ng/ml) is good enough for the confirmation of all immunological analytical results (see Fig. 8-25). Reaction of benzodiazepines with pentafluoropropionic anhydride, analogous to acetylation, is another possibility. However, the commercially available spectrum libraries are often incomplete in the case of these reaction products.

Table 8-14. Benzodiazepines and their metabolites that can be determined in urine after acetylation

RI	Name	Characteristic masses with intensities		
2490	Bromazepam HYAC	318_8	247_{66}	121_{100}
2580	Bromazepam-M (hydroxy-) HYAC	334_{54}	247_{100}	
3000	Diazepam-M (nor-hydroxy-) AC	328_{22}	286_{90}	258_{100}
2245	Diazepam-M (nor-) HYAC	273_{30}	230_{100}	
2950	Flunitrazepam-M (amino-) AC	325_{100}	297_{73}	
2725	Flurazepam-M (hydroxy-ethyl) AC	374_{54}	314_{100}	
3025	Flurazepam-M (di-desethyl-) AC	373_{74}	273_{100}	
2820	Midazolam-M (hydroxy-) AC	383_{28}	310_{100}	
3150	Nitrazepam-M (amino-) AC	293_{100}	265_{75}	
2245	Oxazepam HYAC	273_{30}	230_{100}	
2730	Temazepam AC	342_6	271_{100}	
2245	Temazepam-M HYAC	273_{30}	230_{100}	
2790	Temazepam-M (hydroxy-) AC	342_{16}	300_{61}	272_{100}

8.3.3 Amphetamines

In recent times, no other group of substances has featured more frequently or more intensively in the media than the amphetamines, the so-called "designer" drug "ecstasy" being in the forefront of these reports. A common feature of these substances is their stimulation of the central nervous system, which depends on the release of catecholamines. Therapeutically, substances of this group are used as appetite suppressants, antihypotonics and psychoanaleptics. They can be divided into five groups:

- Amphetamines and methamphetamine
- Phenylethylamines
- Methylenedioxyamphetamines
- Methoxyamphetamines
- Methylenedioxybutanamines.

Amphetamine, methamphetamine and their molecular variants are also known as amphetamines of the first generation, and members of the other four groups as amphetamines of the second generation.

Amphetamines and Methamphetamine

These two substances are derived from the catecholamines or from ephedrine, fenetylline being another member of this group. As they do not contain hydroxyl groups, they are highly lipophilic and therefore easily overcome the blood-urine barrier. Because of their powerful stimulant effect on the central nervous system and their consequent habit-forming potential, they are covered by the narcotics regulations. Their principal effects are feelings of enhanced energy and euphoria. Reduction in the desire for food or sleep eventually leads to a state of exhaustion, and after long use, physical decline, with sympathomimetic effects on the heart and circulation. Chronic abuse sometimes results in psychotic and paranoid states. These substances are mainly consumed by young people and members of the drug scene. Apart from these highly active stimulants, there are also less powerfully acting substances whose importance in drug screening is often underestimated. The group includes ephedrine and its derivatives propylhexedrine and prolintane, which are also thought to have be potentially addictive. Abuse of the appetite suppressants also plays a central role in this connection.

Phenylethylamines

The most important natural substance among the phenylethylamines is mescalin, the substance present in *Lophophora williamsii*, Cactaceae. Synthetic products bought and sold in the drug scene also include 3,4-dimethoxyphenylethylamine (DMPEA) and 4-bromo-2,5-dimethoxyphenylethylamine (BDMPEA). These hallucinogenic substances cause optical and acoustic illusions, which have often been deliberately experienced and then recorded [15]. The resulting psychoses, especially in cases of chronic abuse, have sent consumers into a state of exaggerated self-perception in which they have injured themselves while imagining, for example, that they can fly.

Methylenedioxyamphetamine Derivatives

The most important of these substances are 3,4-methylenedioxyamphetamine (MDA), 3,4-methylenedioxyethylamphetamine (MDE) and 3-methoxy-4,5-methylenedioxyamphetamine (MDMA) (see Section 7.2.6.1). MDMA, first synthesized as an appetite suppressant in 1914, and MDA, which has been commercially available since 1910, are strictly speaking not designer drugs (see Section 8.3.4), but all three are often referred to as such and are bought and sold under the name "ecstasy". A new market for these substances, especially in the techno-scene, has developed and grown at a very high rate in recent years. These amphetamine derivatives of the second generation give a feeling of increased contact and communication, and personal problems are judged optimistically. Because of these effects, these substances are also described as love drugs, and this is reflected in the names used in the drug scene: "Adam" for MDMA and "Eve" for MDE. However, use of these drugs is dangerous. Apart from the effect of stimulating the central nervous system, there are also hallucinogenic effects, especially with MDA. The sympathomimetic effects combined with an increased readiness to expend energy tend increasingly to cause circulatory damage, which can have a fatal outcome due to heart stoppage, apoplexy, circulatory collapse or hyperthermia. Non-fatal side effects, partly caused by toxic metabolites, are also considerable, and can include neurotoxicity, liver damage and psychoses [50].

These important products are not all synthetic. The oral consumption of 5–30 g nutmeg will produce hallucinogenic and entactogenic effects. The two constituents elemicine and myristicine (see Fig. 6-1) are converted by transamination in the liver to 3,4,5-trimethoxyamphetamine and MMDA [15].

Methoxyamphetamine Derivatives

In these amphetamine derivatives, the hallucinogenic effects are predominant, exceeding those of the phenylethylamines many times over. Thus, for example, 2,5-dimethoxy-4-methylamphetamine (DOM) has approximately 100 times the effect of mescalin [11]. Its effects of "serenity, tranquility, and peace" give the product its scene name "STP". All substances, including 3,4,5-trimethoxy-amphetamine (TMA), which is also a metabolite of elemicine (a constituent of nutmeg), and 4-bromo-2,5-dimethoxyamphetamine (DOB), have been almost completely displaced by the methylenedioxyamphetamines and today play an insignificant role in the drug scene.

Methylenedioxybutanamines

The substances of this amphetamine subgroup have an effect similar to that of the methylenedioxyamphetamines. Although these sometimes have a hallucinogenic as well as an entactogenic effect, the former disappears on nitrogen substitution, in analogy to MDMA. These substances are currently regarded as of little importance in the drug scene.

8.3.3.1 Metabolism

Amphetamine, Methamphetamine, Fenetyllin

The main route of metabolic breakdown of the *amphetamines* is oxidative deamination to give phenylacetone, which is further broken down to benzoic acid and excreted as hippuric acid after the latter has been conjugated with glycine. A less important process includes hydroxylation reactions to 4-hydroxyamphetamine and norephedrine. The excretion of amphetamine is very dependent on the pH of the urine. An acidic pH leads to accelerated elimination of amphetamine such that the unchanged fraction excreted can increase from 30% to 75% of the amount taken. Approximately 90% of the amphetamine consumed is excreted under normal pH conditions in 3–4 days in the urine. *Methamphetamine* is to a small extent metabolized to amphetamine. Under normal conditions, most of the dose consumed is eliminated renally within 24 h, partly as the 4-hydroxy derivative but mainly unchanged [44]. *Fenetyllin* is broken down to amphetamine and etofyllin, but is also excreted unchanged in the urine, depending on the dose taken.

Ephedrine, Methylephedrine, Propylhexedrine, Prolintane

The biotransformation of *ephedrine* proceeds via demethylation to norephedrine and via oxidative deamination followed by conjugation, in analogy to the metabolism of amphetamine. Over 90% is eliminated via the kidneys within 24 h, up to 75% unchanged, 8–20% as norephedrine and the rest after oxidative deamination. *Methylephedrine* is 33% unchanged and is 8% excreted as ephedrine. The excretion rate also depends on the pH of the urine. Desmethylpropylhexedrine and 4-hydroxypropylhexedrine, metabolites of *propylhexedrine*, can be determined in the urine [44]. *Prolintane*, a benzylbutylpyrrolidine product, is to some extent changed oxidatively in the liver. This mainly produces hydroxylated metabolites, but also produces compounds with a lactam structure. The unchanged substance can be detected in the urine as well as the metabolic products [51].

Mescalin

Most of the mescalin dose is excreted unchanged in the urine, but considerable amounts of 3,4,5-trimethoxyacetic acid are also present. The other products in urine, mainly N-acetylated O-desmethyl derivatives, are present in smaller quantities.

MDA, MDE, MDMA

The metabolic deactivation of MDE proceeds via dealkylation to MDA and piperonylacetone, leading to the formation of a hippuric acid derivative. As well as these three metabolites, 3,4-dimethoxy- and 4-hydroxy-3-methoxyamphet-amine derivatives can be detected in the urine [10]. An analogous route for MDMA has also been described [52].

The two relevant constituents of nutmeg, elemicine and myristicine, become centrally active substances by transamination in the liver. The effects of the biogenic products 3,4,5-trimethoxyamphetamine and MMDA have been known for a long time. After consumption of these drugs, the urine also contains various methoxypropenylphenols, especially 2,6-dimethoxy-4-(2-propenyl)-phenol (Fig. 8-26).

Methoxyamphetamines

After consumption of *p*-methoxyamphetamine (PMA), more than 80% of the amount of substance taken is excreted within a day via the urine, up to 15% unchanged, >25% in the form of the free 4-hydroxyamphetamine, and ca. 50% as conjugated 4-hydroxyamphetamine. The remainder, in analogy to amphetamine, is further metabolized [44]. In analogy to PMA, TMA undergoes a demethylation to form mono- and di-O-desmethyl derivatives. In methoxyamphetamines, two methoxy groups are located in the *para* position (DOB, DOM), so that demethylation to hydroquinone is also possible [51].

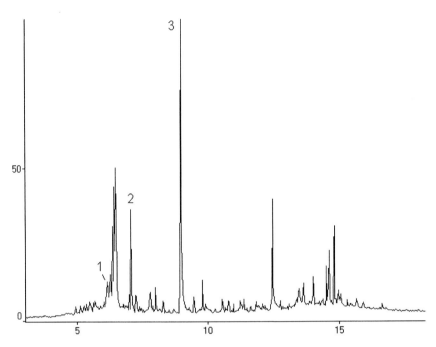

Fig. 8-26. Chromatogram of a urine sample after consumption of large quantities of
nutmeg.
1 = phenol-2-methoxy-5-(1-propenyl); 2 = phenol-2,6-dimethoxypropenyl;
3 = phenol-2,6-dimethoxypropenyl

8.3.3.2 Analysis

Both immunological and chromatographic methods are available for
determining amphetamine consumption from urine. For the immunological
determination, it is important that the sample should not be previously stored
unrefrigerated for a period of days, as amines produced by putrefaction can give
a positive amphetamine result [53]. As can be seen from Table 8-15, the most
important members of this very large and heterogeneous group of substances
can be determined with sufficient sensitivity by the immunological method.
However, in amphetamine determinations there is a considerable danger of false
positive immunoassay results that cannot be confirmed by GC/MS.
Confirmation of an analytical result after prolonged storage can become

problematical if amphetamine derivatives are absorbed from the plastic material of the sample containers [53]. A potential effect on the test result from accompanying medication also cannot be excluded. In cases of this kind, patients who have received high doses of promethazine or perazine have been investigated in our laboratory, but by no means all the samples from the patients who received this high dose of phenothiazines gave a positive amphetamine result. This only occurred when glucuronide cleavage was carried out as described earlier. Figure 8-27 shows the amount of promethazine metabolites in the urine after therapeutic doses had been taken.

Table 8-15. Cut-off values of some amphetamines and their metabolites

Substance	CEDIA®DAU	Emit®d.a.u.[a]	dau-TRAK®	Triage™ 8®
D-Amphetamine	0.3	0.3	0.28	1.0
Ephedrine	60.0	1.0	10.0	500.0
MDA	13.6	35.0	0.4	2.0
MDE	–	–	–	2.5
MDMA	0.43	34.0	–	2.0
D-Methamphetamine	0.3	1.0	0.38	1.0
Prolintan	–	–	–	–
Propylhexedrine	–	2.0	2.0	2.0
Trimethoxy-amphetamine	–	–	100.0	–

[a]Group test for amphetamines; there is also a more specific amphetamine test.

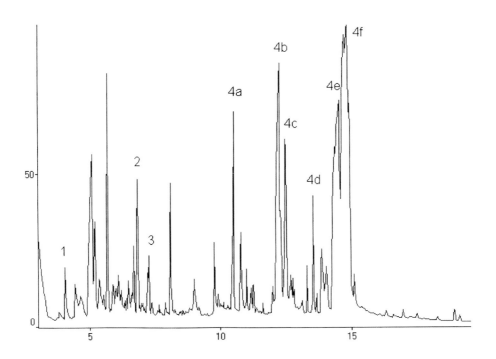

Fig. 8-27. Chromatogram of a urine with promethazine metabolites.
         1 = amphetamine; 2 = MDMA; 3 = MDMA-met. (-desmethylenyl-methyl);
         4a-f = promethazine-met

The connection between the sweetener cyclamate and false positive amphetamine results is discussed in [54].

Chromatographic methods can detect the whole spectrum of amphetamines and their metabolites and enable the various drugs to be distinguished from one another. The successful determination of amphetamines and their derivatives by HPLC/DAD requires a precise knowledge of the retention indices, although the sensitivity is often limited to concentrations above 300 ng/ml. Here, the GD/MS method gives clearly better results. Figure 8-28 illustrates the metabolism of MDE and MDMA using an underivatized sample.

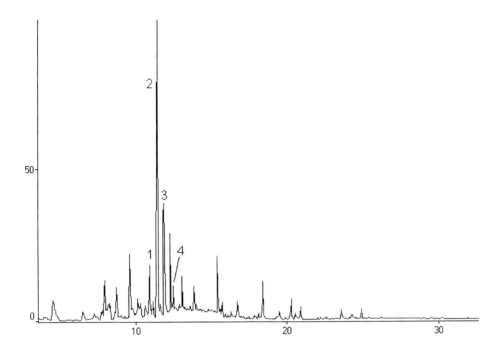

Fig. 8-28. Chromatogram of a urine after consumption of MDE and MDMA.
          1 = MDA; 2 = MDMA; 3 = MDMA-met.; 4 = MDE

In sample preparation, the temperature for evaporation of the solvent should not exceed 60°C, as some amphetamine derivatives are rather volatile. For the evaluation of the mass spectra of amphetamine derivatives, it is preferable to begin the scanning range at 50 Dalton. This avoids the unspecific but very intense base peak at 44 *m/z* and makes it easier to interpret the mass spectra obtained [55]. Advantage is taken of this technique in the mass fragments in Table 8-16.

Table 8-16. Amphetamine derivatives with RI and characteristic mass fragments

RI	Name	Characteristic masses with intensities			
1100	Amphetamine formyl artifact	147_2	91_{12}	56_{100}	
1375	Ephedrine	146_1	131_1	58_{100}	
1370	Ephedrine-M (nor-)	117_2	105_4	57_{100}	
1230	Etilamphetamine	162_2	148_3	72_{100}	
2830	Fenetylline	326_1	250_{100}	207_{23}	
2125	Fenetylline-M (etofylline)	224_{47}	194_{16}	180_{100}	
1250	Fenfluramine	230_1	159_7	72_{100}	
1630	MBDB	207_1	178_2	135_5	72_{100}
1495	MDA	179_7	136_{100}	105_7	
1465	MDA-M (desmethylenyl-methyl-)	181_9	138_{100}	122_{18}	
1560	MDE	207_1	163_1	135_4	72_{100}
1465	MDE-M	181_9	138_{100}	122_{18}	
1495	MDE-M (desethyl-)	179_7	136_{100}	105_7	
1640	MDE-M (desmethylenyl-methyl-)	209_1	137_{12}	122_7	72_{100}
1790	MDMA	193_4	135_{19}	58_{100}	
1810	MDMA-M (desmethylenyl-methyl-)	195_1	137_4	58_{100}	
1690	Mescalin	211_{24}	182_{100}		
1195	Methamphetamine	148_1	134_2	58_{100}	
1885	Methamphetamine-M (hydroxy-)	150_1	135_1	58_{100}	
1430	Methylephedrine	161_1	115_1	72_{100}	
1720	Prolintane	216_1	174_{10}	126_{100}	
2135	Prolintane-M (hydroxy-phenyl-)	232_1	190_3	126_{100}	
1170	Propylhexedrine	155_1	140_3	58_{100}	
1475	Propylhexedrine-M (hydroxy-)	171_3	156_4	58_{100}	

Although many relevant substances can be determined with sufficient sensitivity even without derivatization, it is preferable to acetylate the samples as a standard procedure. After this acetylation, more characteristic mass spectra are obtained (see Figs. 8-7 and 8-8), as the intensity of the molecule peaks increases and fewer fragments are formed (Table 8-17) [55]. The identification of the substances is also improved by the derivatization step. The samples shown in Fig. 8-29 contain small amounts of acetylamphetamine and the derivatization products of some methylenedioxyamphetamines.

Table 8-17. Acetylated amphetamine derivatives that can be determined in urine

RI	Name	Characteristic masses with intensities				
1505	Amphetamine AC	177_4	86_{100}			
1795	Ephedrine 2AC	249_1	148_2	58_{100}		
1675	Etilamphetamine AC	205_1	114_{24}	72_{100}		
1995	MBDB AC	249_1	176_{33}	72_{100}		
1860	MDA AC	221_7	162_{100}			
1930	MDA-M 2AC	265_3	206_{27}	164_{100}		
1985	MDE-AC	249_1	162_{41}			
2080	MDE-M (desmethylenyl-methyl) 2AC	293_1	206_{20}	72_{100}		
2140	MDMA AC	235_1	162_{20}	58_{100}		
2115	MDMA-M (desmethylenyl-methyl) 2AC	279_1	206_{19}	58_{100}		
1575	Methamphetamine AC	191_1	100_{42}	58_{100}		
1995	Methamphetamine-M (hydroxy-) 2AC	249_1	176_6	58_{100}		
1495	Methylephedrine AC	162_1	134_1	72_{100}		
1510	Phentermine AC	191_1	134_6	117_8	100_{59}	58_{100}
2485	Prolintane-M (oxo-di-hydroxy-) 2AC	347_1	279_7	198_{100}		
1570	Propylhexedrine AC	197_{12}	182_{52}	100_{100}		
2390	Tryptamine AC	202_{16}	143_{100}	130_{95}		
2440	Tryptamine 2AC	244_{12}	143_{100}	130_{84}		

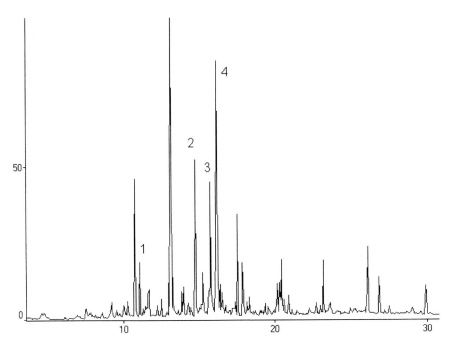

Fig. 8-29. Chromatogram of the urine of an "Ecstasy" user after derivatization.
1 = amphetamine-AC; 2 = MDA-AC; 3 = MDMA-AC; 4 = MDE-AC

8.3.4 Designer Drugs

In order to circumvent the narcotics laws, chemical analogs of opioids or hallucinogens, which are chemical entities of a new "design", are synthesized worldwide in various laboratories. Their pharmacodynamic effects resemble those of the parent substances, which are covered by the narcotics regulations, but their habit-forming potential is often many times greater (see Chapter 6). The possibilities for varying the structure of the model substances in order to develop substances with stronger effects are manifold. Thus, for example, over 1000 derivatives of fentanyl are theoretically possible, and there is a correspondingly large number of substances that show hallucinogenic, entactogenic or centrally stimulating effects.

The drugs are present in dangerously variable concentrations in the materials that are consumed. This dose risk is further increased by the presence in the product of impurities which are side products of the synthesis or starting substances. The designer drugs are divided into five different groups, although the pharmacodynamic profiles of the various substances do not exactly coincide with these groups.

8.3.4.1 Pharmacology and Metabolism

The substances most commonly referred to as designer drugs are derived from *amphetamine*, which has been discussed in Section 8.3.3.

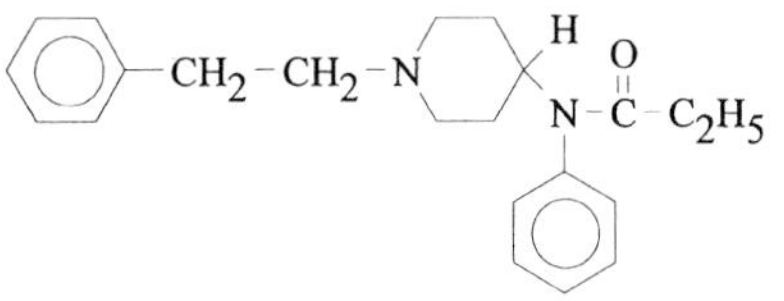

Fentanyl

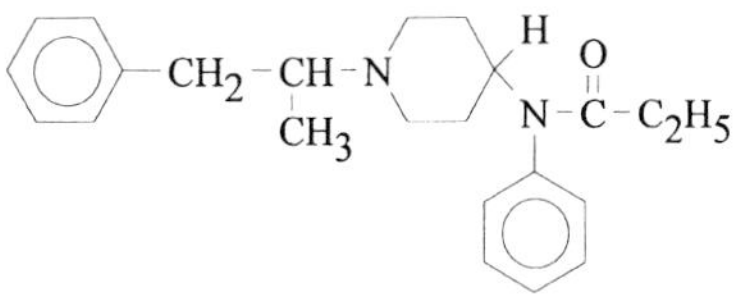

AMF, "China white"

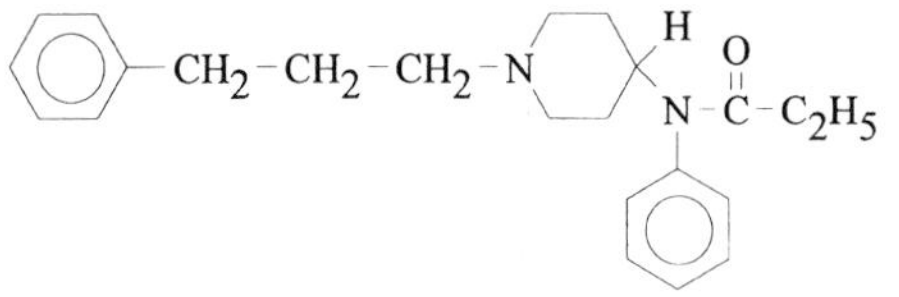

Benzyl-Fentanyl

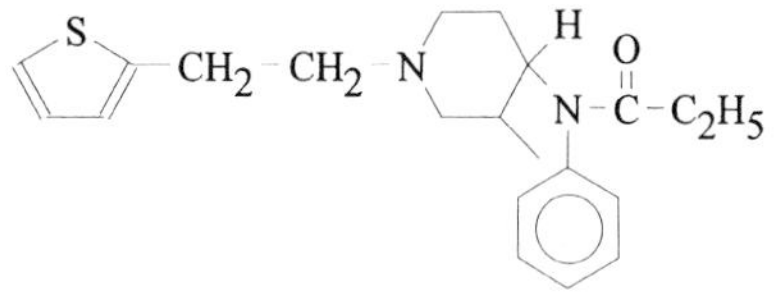

3-Methylthio-Fentanyl

Fig. 8-30. Fentanyl and derivatives not used medicinally

Fentanyl, a short-duration analgesic, is the parent substance of other medicaments which are used in human and veterinary medicine and are permitted for premedication for anesthesia, combination anesthetics and other indications in the field of intensive medicine and anesthesia. All medical preparations of this group have a short-term effect and are therefore of limited interest for the drug scene. Drug abusers therefore prefer molecular variants with longer-term effects, and approximately 30 of these are now available on the illegal market. The fentanyls produce euphoric effects comparable to those of heroin, and they may be used in combination with cocaine. The products that are not used medicinally are therefore regarded in the drug scene as full substitutes for heroin. Fentanyls are many times stronger than morphine in their effects, fentanyl itself by a factor of 100–200, but the depressant effect on respiration is marked, and artificial respiration is necessary in anesthesia. The danger of overdosing and death due to respiratory paralysis is very great compared with heroin, and this is exacerbated by the variable purity of the illegal drugs. The structures of the commonest products are shown in Fig. 8-30 [56].

The metabolism of the illegal substances of the fentanyl group probably involves steps analogous to the deactivation of fentanyl and its derivatives used in medicine. The oxidative N-dealkylation of the piperidine ring leads to the corresponding carboxylic acid and norfentanyl derivatives. Other metabolites are produced depending on which of the many possible substituents are present. The great majority of these substances are excreted renally in the form of inactive metabolites, and the proportion of unchanged fentanyl derivatives in the urine is ca. 10% [44].

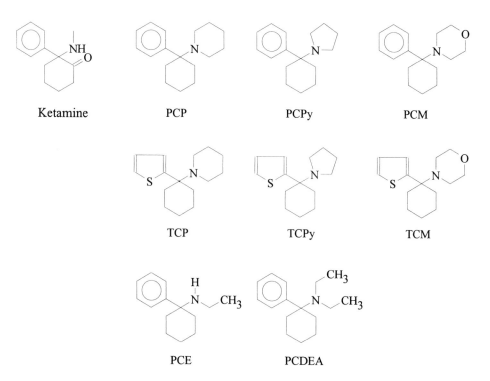

Ketamine   PCP   PCPy   PCM

TCP   TCPy   TCM

PCE   PCDEA

Fig. 8-31. Phencyclidine derivatives

Phencyclidine and its derivatives are analogs of ketamine, which is used therapeutically as an anesthetic. Phencyclidine itself, until the late 1970s, was indicated for this use in the field of veterinary medicine, as in 1965 its use in human medicine had been discontinued because it caused states of excitation and confusion. The symptoms caused by abuse are strongly dependent on the dose, ranging from feelings of euphoria and floating to paranoid states of excitement. Intoxicated patients show much reduced sensitivity to pain. Focal and generalized attacks, which sometimes end in the status epilepticus, are not uncommon. After resorption, phencyclidine, which is highly lipophilic, becomes distributed in the body mainly in the fatty tissues and slowly finds its way into the central nervous system (CNS). On prolonged use, the effects can persist for several days because of the extremely long lifetime of the substance in the CNS. Phencyclidine is hydroxylated in the liver and excreted via the kidneys mainly as glucuronides of 4-hydroxypiperidine or 4-hydroxycyclohexyl derivatives, which then appear in the urine along with the unchanged substance and phenylcyclohexylamine. Figure 8-31 shows some of the many possible phencyclidine derivatives, which differ primarily in the strength of their effects. They include PCE, which has an effect approximately five times as strong as that of phencyclidine (PCP) [57, 58].

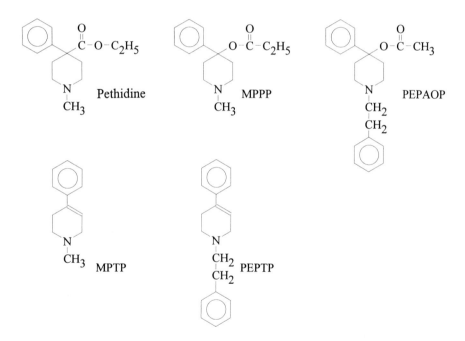

Fig. 8-32. Structural relationships between pethidine and prodines

The model substance of the *prodines* is pethidine (meperidine), a centrally active analgesic with the 4-phenylpiperidinecarboxylic acid structure. The prodines are really inverse pethidine derivatives, whose piperidine system is linked by an ester bond to an aliphatic carboxylic acid. Important members of this as yet not very widespread group include MPPP (1-methyl-4-

phenylpropionoxypiperidine) and PEPAOP (1-phenethyl-4-phenyl-4-acetoxy-piperidine). The effects and side effects resemble those of heroin, although the possible side products of the prodine synthesis are responsible for the much more problematical consequences of the abuse of this drug: MPTP (1-methyl-4-phenyl-1,2,5,6-tetrahydropiridine) and PEPTP, the analogous *N*-phenethyl compound (Fig. 8-32) cause destruction of dopaminergic neurones in the substantia nigra and consequently the Parkinson syndrome [53].

The metabolism of the prodines must be assumed to be hydrolysis of the ester, formation of the corresponding piperidinol derivatives, and N-dealkylation, which is the most important step. Possible hydroxylation products analogous to those formed in the metabolism of phencyclidine may also be produced.

The analogs of the biogenic amine *tryptamine* are mainly found in plants, but sometimes also in the animal kingdom (bufotenine), and are traditionally used, especially in South America, as ritual drugs to produce prophetic visions through hallucinatory experiences. The constituents of the plants used for this are partly derivatives of the neurotransmitter serotonin (5-hydroxytryptamine). Bufotenine, psilocine, psilocybine and 5-methoxydimethyltryptamine are also consumed in Europe to some extent [15]. The effects of these natural drugs, which last for several hours, manifest themselves as euphoria followed by optical and acoustic hallucinations and psychedelic effects. There is a cross-tolerance to the synthetic tryptamines, which produce the same effects. Of these, dimethyltryptamine (DMT), its ethyl homolog (DET) and the more strongly acting 2- and 5-methoxytryptamines are in circulation and are so far not covered by the narcotics laws. Fatalities following high doses of ethyltryptamine (etryptamine or "love pills") with co-consumption of alcohol have been reported.

Tryptamines are metabolized in the liver partly to indoleacetic acid derivatives. Psilocine, the active form of psilocybine, also occurs unchanged in the urine [44].

8.3.4.2 Analysis

With the exception of the amphetamines, "designer drugs" are as yet rather rarely used, as discussed in Chapter 2. Samples are very seldom submitted to laboratories for analysis for these substances. This is especially true for the derivatives of *fentanyl*. Because of the very low doses taken, the dealkylated metabolites can only be determined by GC/MS. Figure 8-9 shows the chromatogram of an analytical sample taken from a patient who had received fentanyl in therapeutic doses.

The dose rates associated with the *phencyclidine* derivatives require the sensitive methods of determination which immunology and chromatography provide. The immunological reagents marketed for the determination of phencyclidine can determine phencyclidine (PCP) itself with the highest sensitivity, followed by the thienyl derivatives and the *N*-alkylamines. Some important cut-off values of immunological tests are listed in Table 8-18.

Table 8-18. Cut-off values of some phencyclidine derivatives in urine

Substance	CEDIA®DAU	Emit®d.a.u.[a]	dau-TRAK®
Phenylcyclidine (PCP)	0.025	0.025	0.1
4-Phenyl-4-piperidinocyclohexanol (4-OH-PCP)	0.032	–	–
1-(1-Phenylcyclohexyl-4-hydroxy-piperidine (PCHP)	1000	–	–
1-(1-Phenylcyclohexyl-)morpholine (PCM)	–	0.09	–
1-(1-Phenylcyclohexyl-)pyrrolidine (PCPy/PHP)	0.025	0.06	–
1-1-(2-Thienyl)-cyclohexyl-morpholine (TCM)	–	0.17	–
1-1-(2-Thienyl)-cyclohexyl-piperidine (TCP)	0.1	0.03	–
1-1-(2-Thienyl)-cyclohexyl-pyrrolidine (TCPy)	–	0.075	–
N,N-Diethyl-1-phenyl-cyclohexylamine (PCDEA)	–	0.35	–
N-Ethyl-1-phenyl-cyclohexylethylamine (PCE)	0.1	–	–

[a] The rapid tests MAHSAN-PCP (MAHSAN Diagnostika) and microLINE Screen for PCP (mölab) can determine phencyclidine (PCP) at a urine concentration of 0.025 µg/ml and TCP at 2.5 µg/ml (MAHSAN or 0.05 µg/ml (microLINE).

The sensitivity of determination by GC/MS exceeds that by immunology, so that concentrations as low as 500 pg/ml can be determined in the urine [59]. The fragmentation of the variously substituted compounds (Table 8-19) enables the consumed PCP derivatives and their metabolites to be distinguished from each other.

Derivatization with acetic anhydride increases the sensitivity towards the secondary amines of the group and the hydroxylated phencyclidine metabolites. The esters and amides so formed are listed in Table 8-20 below.

Table 8-19. Phenylcyclidine derivatives with characteristic masses

RI	Name	Characteristic masses with intensities			
1910	PCP	243_{23}	200_{100}	186_{20}	91_{62}
1960	PCM	245_{29}	202_{100}	168_{14}	91_{86}
1525	PCC	192_{4}	191_{7}	149_{100}	122_{13}
2020	PCIP	258_{20}	99_{100}		
1570	PCDI (=PCDEA)	203_{25}	160_{100}	146_{26}	
1480	PCME	189_{17}	146_{100}	91_{16}	
1625	PCPR	217_{15}	174_{100}	91_{48}	
1975	TCM	251_{9}	165_{62}	97_{100}	
1810	TCPy	235_{19}	165_{34}	97_{100}	
1535	TCDI	209_{18}	165_{42}	97_{100}	

The chromatogram of a sample which contains phencyclidine (PCP) is described in Chapter 13.

Table 8-20. Acetylated phenylcyclidine derivatives with characteristic masses

RI	Name	Characteristic masses with intensities			
1920	PCE-AC	245_{8}	188_{42}	158_{78}	91_{100}
1870	PCM-AC	231_{4}	158_{97}	91_{100}	
1965	PCPR-AC	259_{9}	158_{83}	91_{100}	

Prodines and their metabolites, like the fentanyls, can be determined from body fluids only chromatographically. As no urine samples have so far been available to us for this type of analysis, Fig. 8-33 should be referred to. This shows the results of an analysis after therapeutic consumption of 25 mg pethidine, the model substance of the prodines.

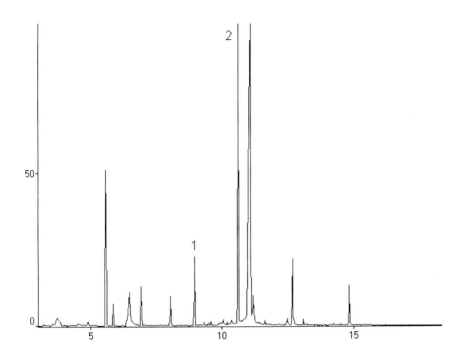

Fig. 8-33. Chromatogram of the urine of a pethidine consumer.
        1 = pethidine; 2 = tramadol

8.3.5 Cocaine

Cocaine is used medicinally as a local anesthetic for the eye, but is traditionally a drug of abuse. Because of the fall in the price of cocaine, its significance as an addictive drug has greatly increased. Usually, cocaine is used by polytoxicomane patients together with opiates, benzodiazepines and cannabis.

8.3.5.1 Metabolism

Only 1–4% of cocaine consumed is excreted unchanged in the urine. In urine with an acidic pH, the unmetabolized fraction can rise to 9%. Cocaine is mainly eliminated as benzoylecgonine and methylecgonine (Fig. 8-34), the metabolism ratios of both metabolites being in the range 30–65%. After cocaine is consumed, the metabolites can be detected in the urine for 2–3 days, depending on the dose. The amount of the unchanged substance after very high doses reaches measurable concentrations.

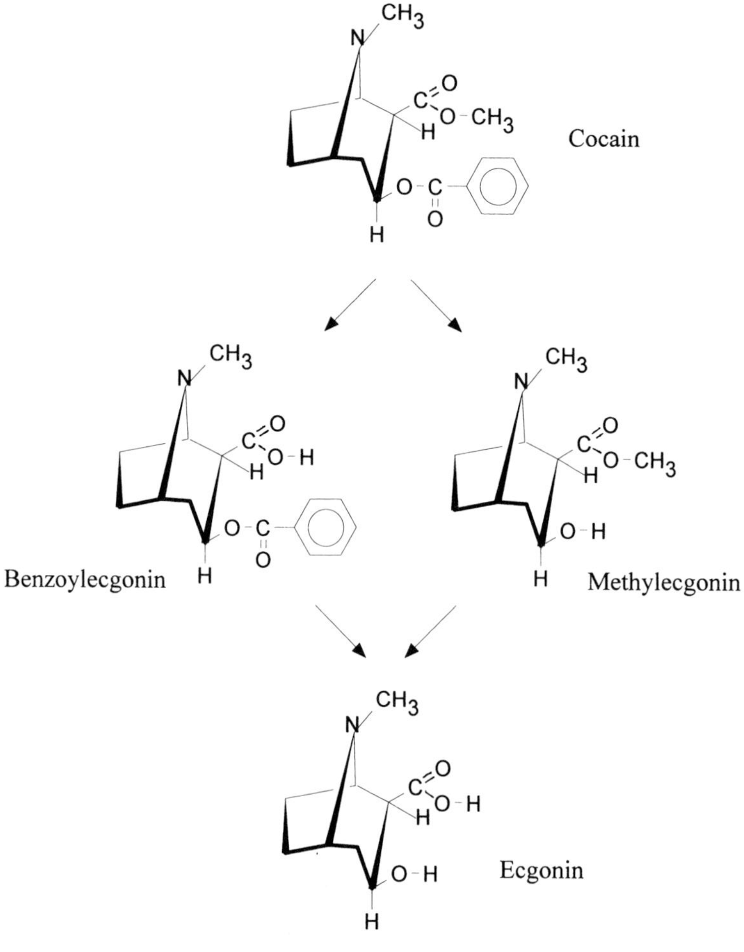

Fig. 8-34. Metabolism of cocaine

8.3.5.2 Analysis

Both immunological and chromatographic methods are available for the determination of cocaine consumption by urine analysis. Table 8-21 gives an overview of the cut-off values of current immunological tests for benzoylecgonine. Methylecgonine also shows appreciable cross-reactivity with the antibodies in some immunological cocaine tests. The reaction behavior towards cocaine itself is less favorable in most tests.

Table 8-21. Cut-off values of cocaine metabolites in urine [µg/ml]

Substance	CEDIA®DAU	Emit®d.a.u.	dau-TRAK®	Triage™ 8®
Benzoylecgonine	0.3	0.3	0.3	0.3
Methylecgonine	>10 000	0.12	–	–
Cocaine	0.547	–	0.3	0.75

The rapid single-step ToxiQUICK-C test (Biomar) detects benzoylecgonine at concentrations above 0.05 µg/ml.

Benzoylecgonine and cocaine can be determined by HPLC/DAD, whereas the molar extinction coefficient of methylecgonine does not permit UV detection with DAD. To confirm a positive immunological result close to the cut-off value, 5 ml urine should be prepared according to the method given in Chapter 7. The determination of cocaine consumption with GC/MS is carried out by measuring methylecgonine after injection of an underivatized urine extract (Table 8-22). As can be seen from Fig. 8-35, an artifact is formed from part of the methylecgonine by splitting off a water molecule. Benzoylecgonine without derivatization is visible in the chromatogram because of the free carboxyl group, but only in cases of high-dose cocaine consumption. To confirm an immunological cocaine determination by GC/MS that is close to the threshold, one of the derivatization methods mentioned below is recommended.

Table 8-22. Cocaine and metabolites that can be determined in urine

RI	Name	Characteristic masses with intensities		
2200	Cocaine	303_{14}	182_{48}	82_{100}
1280	Methylecgonine ($-H_2O$), artifact	181_{31}	152_{100}	
1465	Methylecgonine	199_{18}	82_{100}	
2570	Benzoylecgonine	289_{18}	124_{100}	

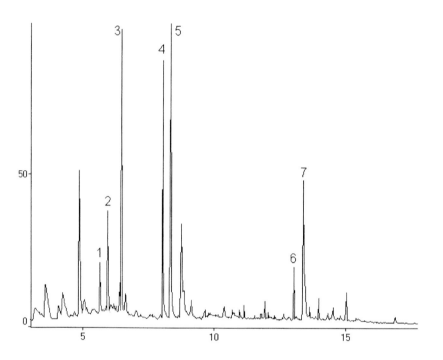

Fig. 8-35. Chromatogram of the sample from a polytoxicomane with cocaine
consumption.
1 = nicotine; 2 = cocaine-met. artifact; 3 = methylecgonine; 4 = cotinine;
5 = meconine; 6 = codeine; 7 = morphine

As an alternative to methylation, treatment of the sample with a mixture of
pentafluoropropionic anhydride (PFP anhydride 100 µl, Aldrich 25, 238-7) and
pentafluoropropanol (PFP propanol 70 µl, Aldrich 25, 747-8) gives derivatives
that are easily measured [60]. The sensitivity of determination of
benzoylecgonine can be so much improved by derivatization that even
determination from serum is possible. Figure 8-17 shows a derivatized sample
in which the benzoylecgonine is present as the PFP derivative.

8.3.6 Chlormethiazole

The range of uses of chlormethiazole extends from the treatment of sleeplessness and mental disturbance of various origins to the prevention of withdrawal symptoms and therapy for the various types of delirium during the treatment of alcoholics. The therapeutic doses range from 200 to 4000 mg/day. The elimination half-life lies between 3 and 7 h. The addictive nature of this drug is so great that it should be prescribed to a patient for the minimum possible time. Its therapeutic range is narrow. Overdosing or co-medication with other substances such as ethanol etc. can lead to a large fall in blood pressure or respiratory failure. Nevertheless, it is often abused by patients suffering alcohol withdrawal symptoms, who consume chlormethiazole and ethanol at the same time. Dose rates can sometimes rise to 20 g/day. The determination of chlormethiazole metabolites in urine is therefore often of toxicological significance.

8.3.6.1 Metabolism

Chlormethiazole is metabolized in the liver at a high rate and is excreted in the urine mainly in the form of inactive metabolites. The proportion of unchanged substance is usually less than 5% of the applied dose. The biological availability therefore increases in patients with impaired liver function. Metabolic deactivation proceeds by stepwise oxidation of the chloroethyl group to 4-methylthiazole-5-acetic acid. Other routes are shown in Fig. 8-36. The main metabolites in urine are the 1- and 2-hydroxyethyl derivatives. Some metabolites are excreted as conjugates [44].

8.3.6.2 Analysis

The determination of this important substance has for a long time been performed exclusively by chromatographic methods, as immunological tests are not available. Although HPLC/DAD can be used, GC/MS is much more sensitive. Methods with a limit of detection of 1 ng/ml have been described for the determination of chlormethiazole from serum [61]. Using the sample preparation methods described in Chapter 7, it is easily possible without derivatization to detect from urine the consumption of a chlormethiazole capsule (192 mg) even on the day after consumption. Figure 8-37 shows the chromatogram of a sample with such a low concentration.

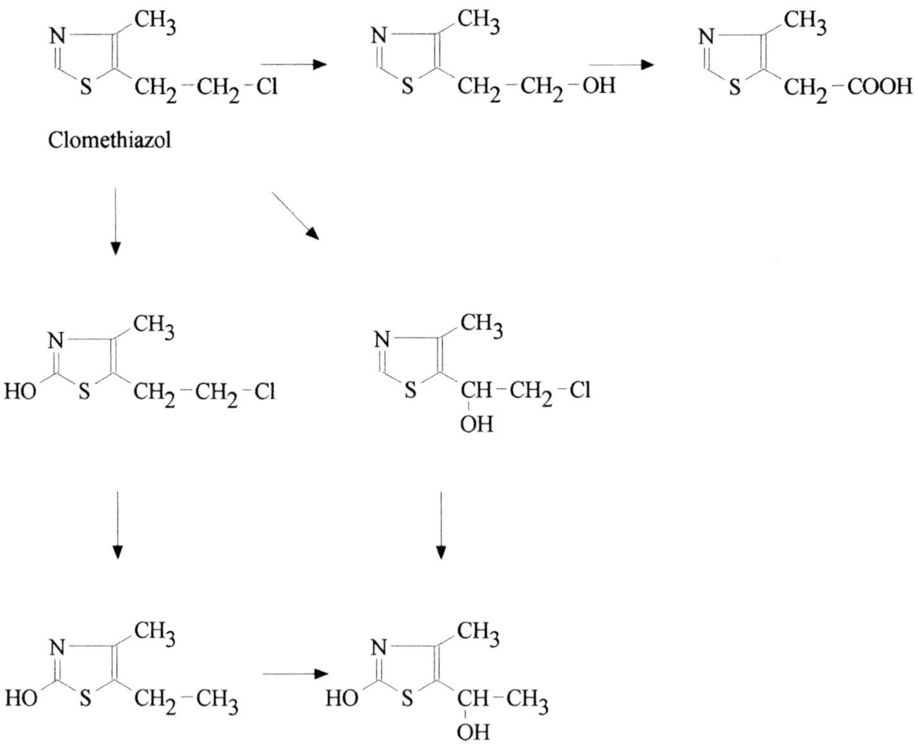

Fig. 8-36. Metabolism of chlormethiazole

Various metabolites can usually be determined in urine samples from patients taking high doses. The chromatogram in Fig. 8-38 shows the urine of a patient who had taken 1152 mg chlormethiazole on the previous day. Several metabolites are detected as well as chlormethiazole.

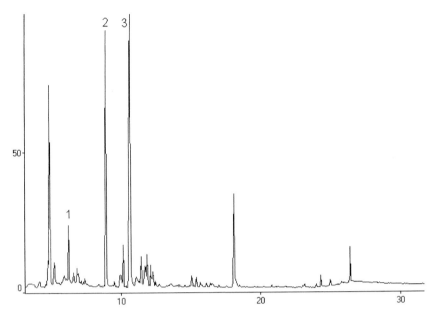

Fig. 8-37. Chromatogram of a urine after consumption of 192 mg chlormethiazole on
the previous day.
1 = chlormethiazole-met. (-2-hydroxy); 2 = cotinine; 3 = caffeine

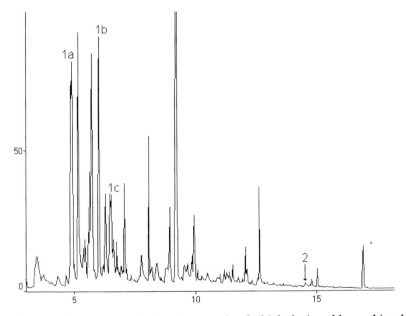

Fig. 8-38. Chromatogram of the urine sample of a high-dosing chlormethiazole patient.
1a-c = chlormethiazole-met.; 2 = 7-aminoflunitrazepam

Table 8-23 gives an overview of the chlormethiazole metabolites that can be detected in urine.

Table 8-23. Metabolites of clomethiazol that can be determined in urine without derivatization

RI	Name	Characteristic masses with intensities	
1160	Clomethiazol-M (deschloro-2-OH)	143_{23}	128_{100}
1380	Clomethiazol-M (deschloro-2-OH-ethyl)	143_{56}	113_{100}
1440	Clomethiazol-M (2-OH)	177_5	128_{100}
1560	Clomethiazol-M (1-OH-ethyl)	177_{27}	124_{100}
1685	Clomethiazol-M (deschloro-di-OH)	159_7	128_{100}

After derivatization with acetic anhydride, the acetylation products listed in Table 8-24 can be identified.

Table 8-24. Acetylation products of clomethiazol that can be determined in urine

RI	Name	Characteristic masses with intensities		
1050	Clomethiazol-M (deschloro-2-OH-ethyl) AC	185_7	143_5	125_{100}
1420	Clomethiazol-M (deschloro-di-OH) AC	183_{19}	128_{100}	
1430	Clomethiazol-M (1-OH-ethyl) AC	219_2	183_{15}	128_{100}
1590	Clomethiazol-M (2-OH) AC	219_3	176_{67}	128_{100}

8.3.7 Methadone

Methadone is sold as the racemate and as the pharmacologically active L-enantiomer. Although only the levorotatory form is permitted in Germany as an analgesic, both forms are used by substitution opiate abusers. When the racemate is used in place of the L-enantiomer, the dose rate must be doubled. For substitution, daily doses of 5–75 mg based on the levo form are prescribed.

8.3.7.1 Metabolism

After consumption of methadone, 20–60% is eliminated within the first 24 h, as much as 30% via the gut. Up to 30% of the methadone present in the urine is unchanged. The extent of excretion in the unchanged form increases with the size of the dose and with decreasing pH of the urine. Approximately half of the amount eliminated via the kidneys is broken down to the desmethyl derivative, almost all of which reacts, undergoing ring closure, to form 2-ethylidene-1,5-dimethyl-3,3-diphenylpyrrolidine (EDDP). Another cyclic demethylation product, 2-ethyl-5-methyl-3,3-diphenyl-1-pyrrolidine (EMDP) accounts for 5–10% of the consumed dose.

8.3.7.2 Analysis

In spite of the incomplete elimination of methadone via the urine, the determination of methadone consumption does not as a rule present any problems. Several immunological and chromatographic methods are available. As EDDP is structurally very different from methadone, the immunological reagents show cross-reactivity to methadone only (Table 8-25).

Table 8-25. Cut-off values for the determination of methadone in urine

Substance	CEDIA®DAU	Emit®d.a.u.[a]	dau-TRAK®	Triage™ 8®
L-Methadone	0.3	0.3	0.3	0.3
Normethadone	–	–	3.0	–
EDDP	500	–	–	–
EMDP	100	–	–	–

In cases of consumption of low daily doses (< 5 mg) of methadone, the use of immunological methods can lead to uncertainty with regard to patient compliance, as, directly after consumption of these low doses, methadone is excreted almost exclusively in the form of the cyclic metabolites. The two instrumental analytical methods GC/MS and HPLC/DAD also determine these metabolites with high sensitivity. GC/MS detects the principal metabolite EDDP as well as methadone itself, as shown in Fig. 8-39. The sensitivity of this analytical procedure for the determination of methadone and EDDP from blood or urine without derivatization is better than 20 ng/ml, i.e. it even extends to the subtherapeutic range [62]. Table 8-26 shows the important masses found in the mass spectrum of methadone and its metabolites. The molecular mass 309 gives only 1% of the intensity of the less specific mass 72. To avoid false interpretations, the retention index or the retention time of the peak should be taken into consideration.

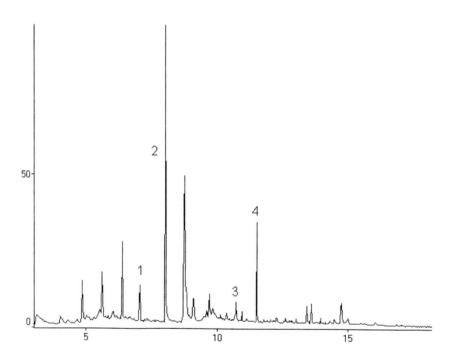

Fig. 8-39. Chromatogram of the urine of a methadone patient with MDE consumption.
1 = MDE; 2 = nicotine; 3 = EDDP; 4 =methadone

Table 8-26. Substances that can be determined in urine after consumption of methadone

RI	Name	Characteristic masses with intensities		
2160	Methadone	309_1	294_3	72_{100}
2040	Methadone M (EDDP)	277_{100}	276_{91}	

8.3.8 Antihistamines

Of the large group of antihistamines, only the active substances diphenhydramine and doxylamine are relevant in drug screening. Both substances are obtainable without prescription in the form of sleeping pills and are consumed by drug-dependent patients in amounts up to ten times the therapeutic daily dose of 50–100 mg. Combination with other medicaments or drugs is not uncommon. The determination of these substances in drug screening is therefore very important, and is often of toxicological significance for the clinician [63].

8.3.8.1 Metabolism

Diphenhydramine is excreted with the urine mainly as diphenylmethoxyacetic acid, partly as the glutamine or glycine conjugate. Other metabolites are formed by hydroxylation and N-dealkylation. A small proportion is also excreted unchanged. Diphenhydramine is eliminated comparatively slowly. Excretion from the organism of 65% of the substance consumed in one dose takes up to 95 h. The excretion of doxylamine is kinetically analogous.

8.3.8.2 Analysis

Immunological methods for the determination of antihistamines are likewise not commercially available. However, both the substances and their metabolites can readily be determined chromatographically using HPLC/DAD or GC/MS using the sample preparation techniques described in Chapter 7. The mass spectra of doxylamine and diphenhydramine are less specific, so that the retention indices must be taken into consideration. However, as shown in Fig. 8-40, it is also possible to determine other metabolites of doxylamine with more specific mass spectra.

The characteristic masses of the two antihistamines are listed in Table 8-27. Derivatization of the samples is not necessary.

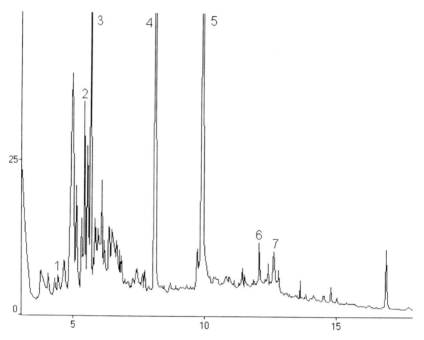

Fig. 8-40. Chromatogram of a doxylamine urine without derivatization.
1 = amphetamine; 2 = chlormethiazole-met.; 3 = chlormethiazole;
4 = cotinine; 5 = doxylamine; 6 = dextorphane; 7 = doxylamine-met

Table 8-27. Antihistamines and their metabolites important in drug screening

RI	Name	Characteristic masses with intensities		
1870	Diphenhydramine	227_1	165_8	58_{100}
1520	Diphenhydramine-M (desamino-OH)	228_9	183_{46}	167_{100}
1890	Diphenhydramine-M (-OH)	213_{11}	183_{25}	58_{100}
1920	Doxylamine	270_1	182_4	58_{100}
1520	Doxylamine-M	183_{57}	182_{100}	
1560	Doxylamine-M (carbinol) –H_2O	181_{35}	180_{100}	

8.3.9 Biperiden

Biperiden, a substance with a parasympatholytic effect, is used in the treatment of Parkinson's disease. It is available on prescription only, the therapeutic dose being 2–12 mg/day. Apart from this use, biperiden is also prescribed for the treatment of extrapyramidal motor disturbances caused by the use of highly active neuroleptic drugs (haloperidol, benperidol etc.). The consumption of 20–30 mg biperiden produces a euphoric effect which can occasionally culminate in a reversible psychosis.

8.3.9.1 Analysis

Only chromatographic methods are available for the determination of biperiden abuse. Using the sample preparation method described in Chapter 7, the hydroxylated metabolites (characteristic masses are given in Table 8-28) can be extracted from the urine. Whereas HPLC/DAD is only capable of detecting high-dose abuse, GC/MS can detect the consumption of therapeutic doses without derivatization.

Table 8-28. Mass spectrometric data for biperiden and metabolites

RI	Name	Characteristic masses with intensities		
2280	Biperiden	311_1	218_{15}	98_{100}
2645	Biperiden-M (-OH)	327_3	218_5	98_{100}
2620	Biperiden-M (-OH)	369_2	269_1	98_{100}

8.3.10 Accompanying Medication

As well as detecting potentially addictive substances, drug screening also provides information about other substances contained in the urine. The use of liquid-liquid extraction enables hormones and cholesterol to be detected. Using both the sample preparation methods, caffeine and other xanthine derivatives can also be extracted. Smokers can be identified by the detection of nicotine and its metabolite cotinine. The determination of these substances in drug screening is usually used only for credibility and quality control. The determination of accompanying medication of the patient is of great importance in this connection. In our laboratory, prescribed medication is taken into consideration when choosing the sample preparation method and the method of analysis, and we are therefore concerned always to provide this information for each drug screening. A false positive result can be caused by accompanying medication, especially if an immunological method of analysis has been used. Therefore, information about accompanying medication must be provided when analysis is by immunological methods, and it must be considered when evaluating the results. In the evaluation of a drug screening, all substances found are quoted. Thus, the person requesting the drug screening receives, in addition to information on addictive substances, information that enables him or her, for example, to make an assessment of either patient compliance or consumption by the patient of other medicaments. For the laboratory, it is useful for purposes of credibility testing (i.e. to be certain of the correct attribution of urine samples to patients) to know about any consumption of accompanying medication. The detection of an unusual type of accompanying medication has led to the discovery that a urine sample under investigation did not originate from the named patient but from another patient from the same clinic. The information is also useful for quality control of the analysis. Thus, the urine from a patient who takes one tablet per day of the sleeping drug flunitrazepam gives, without further effort, useful information on the capability of state-of-the-art GC/MS to determine this important benzodiazepine. A permanent data base of such results is also very useful when advising the physician concerning questions of time required for analysis and limits of detection of particular medicaments. Some medicaments are clinically prescribed in high doses, and, in drug screening by GC/MS, this can lead to the use of an usuitable scale in the representation of the chromatogram, which always shows the largest peak. Therefore, in the evaluation, attention should also be paid to peaks that appear to be very small in comparison to this. Accompanying medication or metabolites can also mask the peaks due to addictive substances. The likelihood of this can assessed by reference to retention times or retention indices, and, if it occurs, it is advisable to look for the relevant substances with the aid of mass fragments. In our laboratory, the following medicaments often occur in the urine in amounts corresponding to high dose rates: carbamazepine, tiapride, promethazine, and other phenothiazines or thioxanthenes and antidepressants (Fig. 8-41). In urine samples of drug patients who have just been received into the clinic, the antidepressant doxepin is often detected.

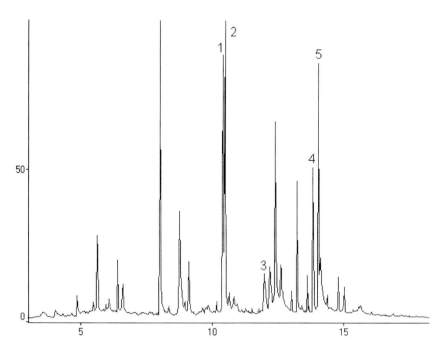

Fig 8-41. Chromatogram of a urine with antidepressant metabolites.
1 and 2 = metamizol-met.; 3 = trimipramine; 4 = trimipramine-met.;
5 = imipramine-met.

8.3.11 Tetrahydrocannabinol

Δ^9-Tetrahydrocannabinol (Δ^9-THC) is a hallucinogen that is often abused to obtain a "high". Consumption is usually by inhalation of the smoke from mixtures with tobacco. Δ^9-THC occurs, together with other cannabinoids, in the resin from the female plant of Indian hemp, *Cannabis sativa*, of which it is the most important constituent. The dried flowering or fruiting tips of the female cannabis plant, also known as marihuana, contain up to 5% Δ^9-THC, while the rods or tablets of the compressed resin, hashish, can contain up to 60%, depending to its origin. The various grades of hashish have names that refer to the country of origin, e.g. "green Turkish", "blond Moroccan", "black Afghan", "white Afghan" or "red Lebanese". HPLC/DAD or GC/MS can be used to precisely characterize the various grades by qualitative and quantitative determination of the cannabinoids present [64].

8.3.11.1 Metabolism

Δ^9-THC is broken down oxidatively in the liver. 11-Nor-Δ^9-THC and other THC derivatives including 11-nor-Δ^9-THC-9-carboxylic acid, together constituting up to 25% of the amount of substance taken, appear in the urine mainly as glucuronides. Elimination via the urine is the less important route, the greater proportion being excreted with the feces. Because of its highly lipophilic properties, Δ^9-THC is stored, mainly in the fatty tissues, for periods sometimes amounting to several weeks, especially after regular consumption. If breakdown of the fatty tissues then occurs due to illness or dieting etc., the Δ^9-THC and its metabolites are released at the same time. Concentrations produced in the urine can then be enough to give positive screening results if a reagent with a sensitivity of 25 ng/ml is used, even though no recent abuse has occurred.

8.3.11.2 Analysis

Up to 65% of the amount consumed is excreted within the first 5 days, and detection of a single consumption of hashish is possible for 12 days by determination of metabolites in the urine. After high-dose or regular consumption of the drug, THC metabolites may often be found in the urine for up to 4 weeks.

Table 8-29. Cut-off values of immunological tests for cannabinoids

Substance	CEDIA®DAU[a]	Emit®d.a.u.[a]	dau-TRAK®	Triage™ 8®
Cannabinol	1.0	–	–	10.0
Cannabidiol	1.0	–	–	–
Δ^9-THC	0.5	–	–	–
11-Nor-Δ^9-THC-9-carboxylic acid	0.05	0.05	0.15	0.05
11-Nor-Δ^8-THC-carboxylic acid	0.04	–	0.1	0.075
8β-11-dihydroxy-Δ^9-THC	0.5	0.1	–	–
8β-Hydroxy-Δ^9-THC	1.0	0.1	–	–
11-Hydroxy-Δ^8-THC	–	0.1	–	–
11-Hydroxy-Δ^9-THC	0.125	0.1	–	–

[a] These values are based on the test for 11-Nor-Δ^9-THC-9-carboxylic acid with a cut-off value of 50 ng/m

Suitable immunological reagents for the detection of cannabis abuse from urine tests are available from several companies. In normal circumstances, the sensitivity of a test corresponds to a cut-off value of 50 ng/ml based on 11-nor-Δ^9-THC-9-carboxylic acid. The cut-off values of some cannabinoids are listed in Table 8-29. Only 11-nor-Δ^9-THC-9-carboxylic acid is relevant to the determination of consumption from urine tests.

The sample preparation method described in Chapter 7 is not suitable for the determination of THC metabolites from urine using GC/MS. THC-carboxylic acid derivatives can only be determined with good sensitivity after cleavage of the glucuronide in alkaline medium after acid extraction and derivatization with dichlorodimethylsilane/toluene [65]. For routine screening from urine, this complex procedure is unsuitable for cost reasons. However, use of this method is essential for the determination of Δ^9-THC, 11-hydroxy-Δ^9-THC and 11-nor-Δ^9-THC-9-carboxylic acid from serum, e.g. to certify that a person is capable of driving a vehicle. The method is also suitable for confirming a positive immunological result. The characteristic masses of the methylation products are listed in Table 8-30.

Table 8-30. Methylation products of THC and some metabolites

RI	Name	Characteristic masses with intensities	
2360	Tetrahydrocannabinol-ME	328_{82}	313_{100}
2620	Nor-Δ^9-THC-COOH-2ME	372_{52}	313_{100}
2840	Hydroxy-nor-Δ^9-THC-COOH-2ME	388_{42}	329_{100}
2860	Oxo-nor-Δ^9-THC-COOH-2ME	386_{55}	327_{100}

References for Part II

[1] Keup, W., *Missbrauchmuster bei Abhängigkeit von Alkohol, Medikamenten und Drogen*: Frühwarnsystem-Daten für die Bundesrepublik Deutschland, Freiburg: Lambertus, 1993.

[2] Landeskriminalamt Nordrhein-Westfalen, *Rauschgiftkriminalität in Nordrhein-Westfalen, Jahresbericht 1993*, Düsseldorf, 1994.

[3] Daunderer, M., *Drogenhandbuch für Klinik und Praxis*, Landsberg, München, Zürich: Ecomed, 1994.

[4] Friessen, K.H., Täschner, K.-L., *Fortschr. Neurol. Psychiat.*, **1991**, 164–169.

[5] Keup, W., *Deutsche Apotheker Zeitung* **1995**, *135*, 129–131.

[6] Sahihi, A., *Suchtreport* **1990**, *5*, 58.

[7] Burian, W., Jedlicka, S., *Wiener Zeitschr. f. Suchtforschung* **1979**, *3*, 13–17.

[8] Alsen-Hinrichs, C., *Suchtmittel – Genuss-, Rausch-, Arznei- und Dopingmittel*, 5, Aufl., Kiel: Neuland, 1993.

[9] Müller, W.E., *Deutsche Apotheker Zeitung* **1991**, *131*, 885–890 and 942–946.

[10] Kovar, K.-A., *Pharmazeutische Zeitung* **1995**, *140*, 1843–1849.

[11] Forsthäusler, D., *Kriminalistik* **1993**, *8*, 533–558.

[12] Sahihi, A., *Designer-Drogen*, 2. Aufl., Weinheim; Basel: Beltz, 1991.

[13] Kovar, K.-A., *Deutsche Apotheker Zeitung* **1992**, *132*, 2302–2305.

[14] Hoffmann, K.-D., *Pharmazeutische Zeitung* **1995**, *140*, 36–42.

[15] Martinez, D., *Rauschdrogen und Stimulantien*, Leipzig; Jena; Berlin: Urania, 1994.

[16] BKA, *Rauschgiftjahresbericht 1993*, Wiesbaden 1994.

[17] Keup, W., *Pharmazeutische Zeitung* **1995**, *140*, 3741–3746.

[18] Katung, W., Klug, E., *Med. aktuell* **1991**, *17*, 654–655.

[19] Thomasius, R., *Lösungsmittelmissbrauch bei Kindern und Jugendlichen*, Freiburg: Lambertus, 1988.

[20] Van Horne, K.C., *Handbuch zur Festphasenextraktion*, ICT GmbH, 65903 Frankfurt, 1993.

[21] Chen, X.-H., Dissertation, Rijksuniversiteit, Groningen 1993.

[22] Gerhards, P., Szigan, J., *Labor Praxis*, **1994**, *10*, 45–50.

[23] Bons, U., Sawazki, J., 1. Viersener HPLC-DAD / GC-MS – Workshop zum Thema Drogenscreening, Viersen.

[24] Iten, P. X., *Fahren unter Drogen- oder Medikamenteneinfluss*, Institut für Rechtsmedizin der Universität Zürich, 1996.

[25] Shimadzu Application Note GC/MS 28, Determination by GC/MS of Drugs of Abuse Mentioned in § 24a Traffic Law; P. Gerhards, Shimadzu Deutschland GmbH, Duisburg, July 1996.

[26] Hügel, H., Junge, W.K., *Deutsches Betäubungsmittelrecht*, 7th edn. with 2nd suppl., Stuttgart: Deutscher Apotheker Verlag, 1995.

[27] Gibitz, H.J., Geldmacher-von Mallinckrodt, M., *Klinisch-toxikologische Analytik bei akuten Vergiftungen und Drogenmissbrauch,* Weinheim: VCH Verlagsgesellschaft, 1989.

[28] Gelsmacher-von Mallinckrodt, M., Enders, P., Bösche, J., Harzer, K., Machata, G., von Meyer, L., Riesselmann, B., *Empfehlungen zum Nachweis von Suchtmitteln im Urin,* Weinheim: VCH Verlagsgesellschaft, 1985.

[29] Kauert, G., Röhrich, J., Schmidt, K., *Toxichem + Krimtech* **1995**, *62*, 14–20. See also Täschner, K.-L., *Deutsche Apotheker Zeitung* **1994**, *134*, 3299–3305 and Iten, P. X., *Fahren unter Drogen- oder Medikamenteneinfluss* 1994, Zürich: Institut für Rechtsmedizin.

[30] Arnold, W., *Akt. Dermatologie* 1989, *15*, 223–229.

[31] Schütz, H., Ahrens, B., Erdmann, R., Rochholz, G., *Pharmazie in unserer Zeit* **1993**, *22*, 65–77.

[32] Maureau, A.P., *MTA* **1996**, *11*, 4–11.

[33] Jopp, C., Diplomarbeit Fachhochschule Krefeld 1996.

[34] Käferstein, H., Sticht, G., *Opiatnachweis im Harn,* Weinheim: VCH Verlagsgesellschaft, 1993.

[35] Käferstein, H., Sticht, G., *Labor Medizin* **1992**, *15*, 459–463.

[36] Arnold, W., Sachs, H., *Fresenius J. Anal. Chem.* **1992, *342*,** 787–790.

[37] Bons, U., *Deutsche Apotheker Zeitung* **1994**, *134*, 3293–3296.

[38] Comprehensive tables with retention indices have been published in the following books: Pfleger, K., Maurer, H.H., Weber, A., *Mass Spectral and GC Data of Drugs, Poisons, Pesticides, Pollutants and their Metabolites,* Weinheim; New York; Basel; Cambridge: VCH Verlagsgesellschaft, 1992. Ardrey, R.E., de Zeeuw, R.A., Finkle, B.S., Franke, J.P., Moffatt, A.C., Möller, M.R., Müller, R.K., *Gas-Chromatographic Retention Indices of Toxicologically Relevant Substances on SE-30 or OV-1,* Weinheim: VCH Verlagsgesellschaft, 1985. Unless stated otherwise, the information on retention indices and mass spectra is from the first-named source.

[39] Schomberg, G., *Gaschromatographie,* Weinheim: VCH Verlagsgesellschaft, 1987. See also the English translation *Gas Chromatography* published in 1990.

[40] Budzikiewicz, H., *Massenspektrometrie*, Weinheim New York, Basel, Cambridge: 1992.

[41] For matching of spectra in our laboratory we use our own libraries as well as the libraries NIST 53, PMW-TOX2 from the sofware CLASS 5000.

[42] Only in the case of tilidine are aromatic and amino groups linked through two C atoms.

[43] Cohen, A.J., Klett, C.J., Ling, W., *Drug Alcohol Dependence* **1983, *12*,** 167–172.

[44] Moffat, A.C., Jackson, J.V., Moss, M.S., Widdop, B., *Clarke's Isolation and Identification of Drugs in Pharmaceuticals, Body Fluids, and Post-mortem Material,* London: The Pharmaceutical Press, 1986.

[45] Penning, R. et al., *Dt. Ärzteblatt* **1993**, *90*, 345–346, Friessem, D.H., Täschner, K.-L., *Fortschr. Neurol. Psychiat.* **1991**, *59*, 164–165.

[46] Albinus, M., *Analgetika und Schmerztherapie*, Wiss. Verl.-Ges., Stuttgart 1988.

[47] Käferstein, H., Sticht, G., *Opiatnachweis im Harn*, Weinheim: VCH Verlagsgesellschaft, 1993.

[48] Burian, W., Jedlicka, S., *Wiener Zeitschr. für Suchtforschung* **1979**, *2*, 13–17.

[49] Weatherall, R., *Journal of Analytical Toxicology* **1994**, *18*, 382–384.

[50] Kovar, K.A., Rösch, C., Rupp, A., Hermle, L., *Pharmazie in unserer Zeit*, **1990**, *19*, 99–107.

[51] Rücker, G., Neugebauer, M., Zhong, D., *Arch. Pharm.* **1992**, *325*, 47–52.

[52] Osterloh, J.D., *Abused Drugs Monograph Series: Amphetamines*, Abbott, 1994 Irving/Texas.

[53] Hofer, R., *Toxichem + Krimtech* **1993**, *60*, 83–85.

[54] Martz, W. et al., *Clin. Chem.*, **1991** *37*, 2016–2017.

[55] Rösch, C., Kovar, K.A., *Pharmazie in unserer Zeit*, **1990**, *19*, 211-221.

[56] Rübsamen, K., *Toxichem + Krimtech* **1987**, *54*, 13.

[57] Daunderer, M., *Drogen-Handbuch für Klinik und Praxis*, Landsberg; München; Zürich: Ecomed, **1990**, *III-3.3*, 1-12.

[58] Haerer, M., Kovar, K.A., *Pharmazie in unserer Zeit*, **1994**, *23*, 52–61.

[59] Berberich, D., Uhrich, M., Tillmanns, U., *GIT Fachz. Lab.* **1990**, 629-632.

[60] Möller, M.R., Bregel, D., Hartung, M., Warth, S., *Toxichem + Krimtech* **1995**, *62*, 28–30.

[61] Jostell, K.G. et al., *Acta Pharmac.Ttox.* **1978**, *43*, 180–189.

[62] Kang, G.I., Abbott, F.S., *J. Chromatography* **1982**, 231.

[63] Daldrup, Th., *Toxichem + Krimtech* **1994**, *61*, 9.

[64] Kovar, K.A., Fanzutti, R., *Pharmazie in unserer Zeit* **1987**, *16*, 33–38.

[65] Daldrup, Th., *Toxichem + Krimtech* **1995**, *62*, 21–27.

Part III Environmentally Hazardous Substances in Working Medicine

9. Quantitative Determination of Pentachlorophenol and Lindane in Blood

P. Gerhards, J. Szigan, A. Wertmann

9.1 Pentachlorophenol

The pesticide pentachlorophenol (PCP) is a typical example of a toxic chlorinated compound that is harmful to the environment. Because large amounts of this substance were produced and used in the past and because of its relatively high volatility and poor biological degradability, it is one of the chemicals that is of ubiquitous occurrence.

PCP has good fungicidal and insecticidal properties and was mainly used as an active substance in industrial and domestic wood protection agents. PCP-containing preparations were also used for impregnating leather and textiles. Since October 1993, chemical prohibition regulations and hazardous substances regulations have applied to PCP, covering its manufacture, marketing and use [1]. As its half-life period in wood is around 6 years, emissions of PCP can occur over many years after application [2]. However, it is still produced and used in several countries. It also has an impact on our environment due to the importation of PCP-treated products (leather, textiles, wood) [3]. Toxicity data and levels of PCP are given in Table 9-1.

PCP enters the human organism via the skin, the lungs and the gastrointestinal tract, and, while the symptoms of acute intoxication with this substance are described in the literature, the long-term harmful effects of prolonged exposure in the home both to PCP and to impurity levels of dioxin, a possible by-product of PCP manufacture, are currently attracting increasing attention and discussion.

Like PCP, the insecticide lindane found a broad spectrum of application in wood preservation and for the treatment of pests affecting soil and forests. Because of the relatively high volatility and slow rate of decomposition of lindane in the environment, this substance, like PCP, also occurs ubiquitously. Table 9-2 gives toxicity data and levels of occurrence of lindane in the environment.

Table 9-1. Normal levels of PCP and toxicity data

Medium	Normal level	Toxicity data
Serum	< 25 µg [4]	ECW[a] 1000 µg/l at 50 µg/m³ in air
Urine	< 25 µg [4]	ECW 300 µg/l at 50 µg/m³ in air
Drinking water	10 µg/l (WHO[b] limit)	
Air at workplace	MAK[c] not available	1990 classification according to III A2 (DFG) (causes cancer in animals)

[a] ECW: Exposure equivalent for carcinogenic substances at the workplace: relationship between concentration in air and concentration of the substance or its metabolites in the biological material
[b] WHO: World Health Organization
[c] MAK: Maximum concentration at the workplace

Table 9-2. Normal levels of lindane and toxicity data

Medium	Normal level	Toxicity data
EDTA-blood	< 25 µg [4]	BAT[a]: 20 µg/l
Drinking water	0.1 µg/l	
Food	2 mg/kg in meat [6]	
	0.7 mg/kg in poultry [6]	
	0.2 mg/kg in milk products [6]	
Air at workplace	MAK 0.5 mg/m³	

[a]BAT: Biological tolerance figure at the workplace

Again like PCP, lindane enters the body via the lungs, skin and gastrointestinal tract, and special attention must be paid to illnesses that can be caused by exposure over many years to both the compound itself and to the toxic impurities produced during its manufacture.

9.2 Analytical Techniques

We discuss here techniques used by us for the analysis of PCP and lindane by gas chromatography with mass spectrometric detection. A GC-17A gas chromatograph with advanced flow control, split-splitless injector and a QP-5000 mass spectrometer were used, and the data were evaluated with CLASS-5000 software. Library searches were based on the NIST 75.000 and Pfleger-Maurer-Weber libraries for drugs and pesticides.

Figure 9-1 shows the equipment parameters, and Table 9-3 lists the SIM parameters used for lindane and PCP.

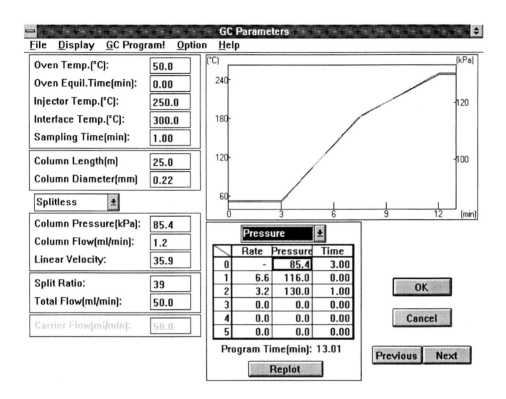

Fig. 9-1. Equipment parameters for GC analysis

The separation was performed on an HT-8 column supplied by the company SGE: 25 m, 0.22 mm ID, with a 0.25 µm film. It was temperature programmed as follows: 50°C/3 min, 30°C/min to 180°C, 15°C/min to 300°C, and a holding time of 1 min.

Table 9-3. SIM parameters for lindane and PCP

Component	Masses used	Sampling rate (s)	Gain
Tribromophenol[a]	331.60; 329.65	0.5	2.50
PCP	265.70; 267.65	0.5	2.50
ε-HCH[a]	180.90; 182.95	0.5	2.50
γ-HCH	180.90; 182.95	0.5	2.50

[a] Internal standard

9.3 Sample Preparation

In the quantitative determination of lindane and PCP from the matrix blood, the sample preparation process is an essential step for ensuring that the results will be of good quality. In this process, interfering substances contained in the matrix are removed, and the substances to be determined are converted into a chemical form that enables them to be measured by GC/MS.

PCP and lindane are chemically different, and appropriate methods of sample preparation must be used. Lindane can be measured by GC without derivatization, but PCP contains a polar functional group and must first be converted into a form in which it readily vaporizes in the injector. For this, a derivatization step is necessary.

In the experimental techniques described below, various sample preparations are discussed. The special problems that occur in the determination of PCP by GC are also discussed, together with some possible solutions.

Blood is a matrix that consists mainly of water, fats and protein. The pollutants under investigation must be extracted quantitatively and reproducibly from this mixture. Criteria for this extraction include the complete separation of the matrix combined with a high recovery rate of the components to be determined.

In the extraction from blood, the bonding of protein to the substance to be extracted and the high viscosity of the blood have a negative influence on the recovery rate. For this reason it is often stated in the literature that the blood should be hemolyzed. The problems caused by this step are described below.

Conventional liquid-liquid extraction for the determination of lindane has been used very frequently up to the present time. However, this classical method of sample preparation has disadvantages, e.g. emulsion formation, poor phase separation, high consumption of solvent, and the labor-intensive nature of the process and its unsuitability for automation. It was important to find ways of reducing the amount of solvent used, especially from the point of view of environmental pollution, but also to give a cost-optimized analysis.

9.3.1 Principle of Sample Preparation for Lindane

In view of the problems described above, various carrier materials for solid phase extraction were tested. Criteria used included the injection of the sample, the distribution on the carrier and the consequent reproducibility of the results. The following carrier materials were tested:

- RP-18, 40 μm particle size for HPLC, powder
- RP-8, 40 μm particle size for HPLC, powder
- Microspheres for the solid phase extraction
- Extrelute®.

The first three carrier materials were found to be unsuitable. The grain size of the commercially available materials was too small. The material adhered to the blood, so that separation became impossible. The small surface area of the materials prevented distribution on the carrier from occurring. Both whole blood and hemolyzed blood were tested.

Extrelute® consists of coarsely porous diatomite with a granular structure and a high pore volume. This material is supplied in pre-packed columns containing various amounts of material.

The sample is applied to the carrier material and becomes distributed on the inert carrier matrix as a thin film, so acting as the "stationary phase". It is then eluted with an organic solvent immiscible with water. All lipophilic substances such as lindane are extracted from the aqueous phase by the organic phase. The matrix remains on the carrier.

The reagents used were:
- ε-HCH, internal standard, Promochem, certified standard
- γ-HCH, Promochem, certified standard
- Extrelute®, 3 ml, Merck
- Isooctane, Merck, GC grade
- Concentrated sulfuric acid, ultra-clean.

The sample preparation scheme for lindane was as shown in Fig. 9-2.

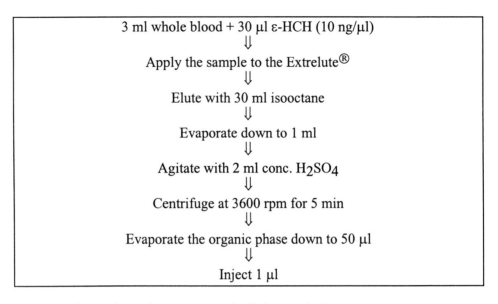

Fig. 9-2 Scheme of sample pretreatment for lindane analysis

This sample preparation scheme is the result of a number of experiments, in which, for example, the amount and nature of the solvent was varied. The criterion of quality was the recovery rate. The investigation was also carried out using whole blood and hemolyzed whole blood. The use of hemolyzed blood led to problems in the sample preparation. Here, recovery rates of only < 20% could be achieved. In the following discussion, only EDTA (ethylenediamine-tetraacetic acid)-whole blood was used.

Table 9-4 shows the solvent and the amount of solvent used in the extraction, the concentration of lindane and the resulting recovery rates.

Table 9-4. Extraction parameters for lindane analysis in blood

Solvent	Amount	Concentration	Recovery rate (%)
Toluene : hexane (20 : 80)	15 ml	100 pg	< 20
Isooctane : cyclohexane (1 :1)	15 ml	100 pg	25
Cyclohexane	15 ml	100 pg	20
Hexane	15 ml	100 pg	54
Isooctane	15 ml	100 pg	57
Isooctane	30 ml	100 pg	95
Isooctane	15 ml	10 pg	96

Table 9-4 shows that the recovery rate is dependent both on the solvent and on the amount of solvent used. For a low concentration, in our example 10 pg lindane/3 ml blood,   15 ml solvent is quite sufficient. However, if 100 pg lindane is present, the resulting recovery rate is poorer, and a larger amount of solvent is therefore used. As the concentration is unknown in real samples, 30 ml solvent should always be used in order to be able to reproducibly determine higher concentrations also.

Figure 9-3 shows the calibration line of lindane. This line was used to determine the recovery rates. For this, the peak area of a standard was determined. The standard was then added in known concentration to a sample of pooled blood and the sample preparation method described above was applied. The recovery rate can be found from the peak area of the standard and that of the blood with the added standard. Each calibration point shown in Fig. 9-3 represents the mean of five individual measurements from the matrix blood.

Each point on the graph corresponds to 5 individual measurements

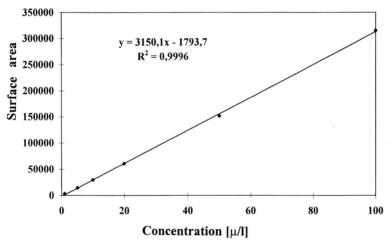

Fig. 9-3. Calibration line for lindane from blood

Figure 9-4 shows full-scan and SIM spectra of 1 pg lindane added to blood and subjected to sample preparation using the method described above.

In the measurement of real samples, the method of the internal standard was used for quantification. This method is especially suitable here, as various matrix effects occur with blood. The internal standard is subjected to the same conditions as the component to be measured. The internal standard used was ε-HCH, as this is chemically very similar to γ-HCH but does not occur in nature.

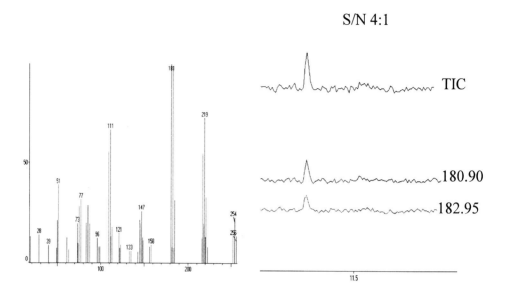

Fig. 9-4.   SIM run of 1 pg lindane added to blood and the mass spectrum of lindane
generated by the full-scan run.

The various compositions of the blood led to problems from interfering
background masses.

Figure 9-5 shows the chromatogram of a real sample with and without
addition of concentrated sulfuric acid, the same sample to which a known
amount of standard was added, and a positive real sample treated with sulfuric
acid.

From the ratio of the masses to each other in the analysis of the real sample,
it can be deduced that a coelution has occurred. This can be caused by fatty
constituents or by breakdown products of other pesticides. For this reason, it is
essential to agitate the sample with concentrated sulfuric acid. As the figures
show, the standard is not attacked by the sulfuric acid, whereas the interfering
components are eliminated. This is also illustrated by the recovery rates for 30
ml isooctane shown in Table 9-4. An attempt was made to purify the samples
with silica gel instead of by using sulfuric acid, but satisfactory results could not
be achieved.

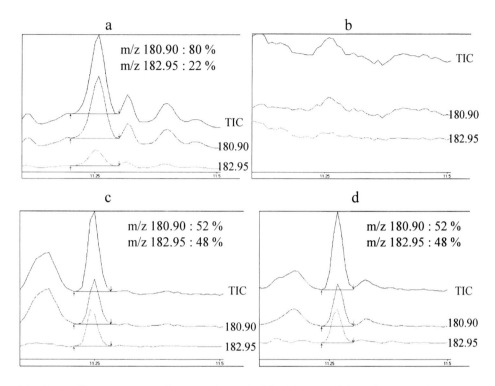

Fig. 9-5. Chromatogram of an actual sample (a) without addition of conc. H_2SO_4,
(b) with addition of conc. H_2SO_4, (c) with addition of the standard. A positive
actual sample with conc. H_2SO_4 is shown in (d).

9.3.2 Principle of Sample Preparation for PCP

The reagents used were as follows:

- 2,4,6-Tribromophenol (TBP), internal standard, Promochem, certified standard
- Pentachlorophenol (PCP), Promochem, certified standard
- Toluene, Merck, GC grade
- Isooctane, Merck, GC grade
- Acetic anhydride, Merck, p.a. grade
- Potassium hydrogen sulfate, Merck, p.a. grade
- Potassium carbonate, Merck, p.a. grade.

The sample preparation was carried out in accordance with the scheme in Fig. 9-6 [4].

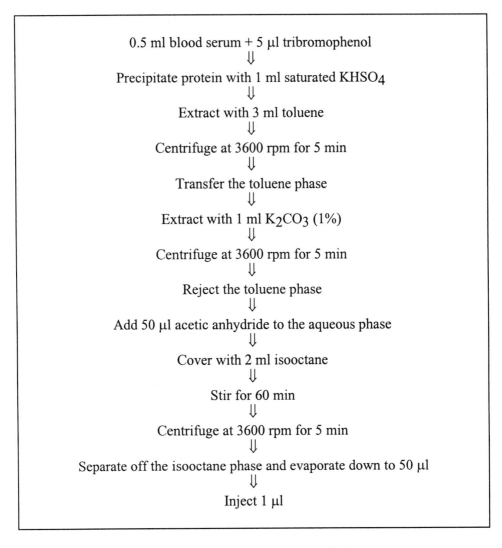

0.5 ml blood serum + 5 μl tribromophenol
⇓
Precipitate protein with 1 ml saturated KHSO$_4$
⇓
Extract with 3 ml toluene
⇓
Centrifuge at 3600 rpm for 5 min
⇓
Transfer the toluene phase
⇓
Extract with 1 ml K$_2$CO$_3$ (1%)
⇓
Centrifuge at 3600 rpm for 5 min
⇓
Reject the toluene phase
⇓
Add 50 μl acetic anhydride to the aqueous phase
⇓
Cover with 2 ml isooctane
⇓
Stir for 60 min
⇓
Centrifuge at 3600 rpm for 5 min
⇓
Separate off the isooctane phase and evaporate down to 50 μl
⇓
Inject 1 μl

Fig. 9-6. Scheme of sample pretreatment for PCP analysis

This sample preparation method (Method 1) is the result of several experiments. An attempt was also made to extract the deproteinated sample with alkali (Method 2) without previously extracting with toluene. As a comparison, an experiment was carried out in which the toluene phase was evaporated to dryness and then acetylated (Method 3). The results are given in Table 9-5. The criterion was again the recovery rate.

Table 9-5. Methods of sample preparation with corresponding recovery rates

Sample preparation method	Recovery rate (%)
Method 1	85
Method 2	< 50
Method 3	63

The differences in the recovery rates are due firstly to the polarity of PCP and secondly to its volatility. It is therefore important to cover the phase with isooctane
during the acetylation stage. Also, the CO_2 formed drives the PCP out of the sample.

Because of the polar hydroxyl group on PCP, acetylation is an important step to enable this component to be measured by gas chromatography. Figure 9-7 shows the MS spectrum, the chromatogram, and the structural formula of PCP. If it is not acetylated, PCP is so polar that large amounts are adsorbed in the injector. To obtain good reproducibility, it is also important to silanize both the insert and the glass wool in the injector. If this is not carried out, losses can occur.

Figure 9-8 shows the calibration curve for PCP which was used to determine the recovery rates. This curve was obtained by a method analogous to that used to produce the curve for lindane.

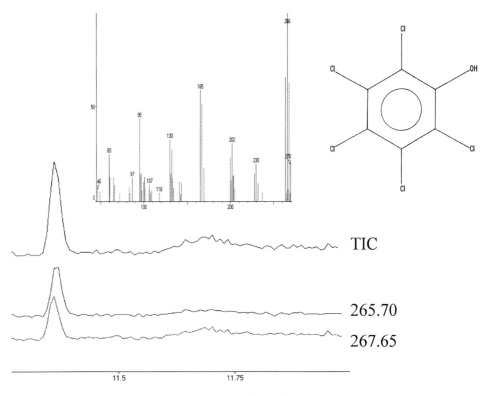

Fig. 9-7. SIM run of 5 pg PCP added to blood and the mass spectrum of PCP
          generated by the full-scan run

Each point on the graph corresponds to 5 individual measurements

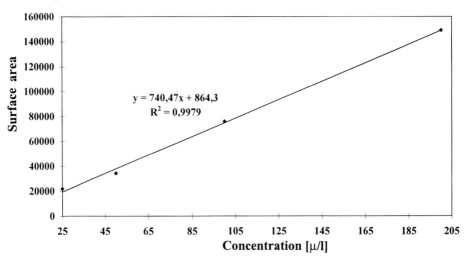

Fig. 9-8. Calibration line for PCP from serum

9.4 Measurement of Real Samples

For the quantitative determination of PCP from serum, the method based on an internal standard was used. The internal standard was tribromophenol.

Figure 9-9 shows the SIM curve of a real sample. Here also, the ratios of the masses to each other give an important indication of possible coelutions.

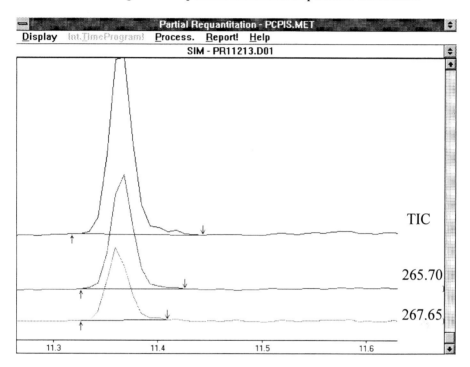

Fig. 9-9.   SIM run of an actual sample.

9.5 Summary

GC/MS offers the user a reliable and sensitive method for the analysis of PCP and lindane. The example of the real sample contaminated with lindane shows clearly that an independent confirmation of the results is necessary to avoid false positive results.

GC/MS provides high sensitivity, especially with SIM. The legally required limits can be achieved by the use of suitable concentration steps during the sample preparation.

It should again be mentioned here that the GC/MS technique can be fully automated. This gives a considerable cost saving.

As GC/MS systems can now be obtained at very advantageous prices, this reliable analytical technique will in the future continue to be introduced more and more into clinical laboratories.

10. Headspace Gas Chromatography in Clinical Chemistry

P. Gerhards, J. Szigan

10.1 Introduction

The principle of the method in which a gas chromatographic column is fed by a static headspace is fundamentally as follows. The sample is introduced into a vial which is clamped with a septum and a cap. The sample is then conditioned for a predetermined period at a predetermined temperature. This leads to an equilibrium state between the components in the liquid phase and those in the gas phase. This equilibrium is dependent on temperature and time, and, as most components of medical significance are volatile, they are mainly to be found in the headspace after the conditioning period. Next, an aliquot part is taken from the headspace by a gas-tight syringe which has been brought up to the correct temperature, and this is then automatically transferred into the gas chromatograph, where separation of the components takes place. Depending on the nature of the components, detection is by FID or ECD. The whole process is automated, enabling up to 40 samples to be analyzed in one series. This process is illustrated graphically in Fig. 10-1.

An important factor in headspace GC is the effect of the matrix on the concentration in the gas phase of the substances under investigation. In the medical field, determinations are usually from blood or urine. As these are organic matrices, the conditioning time and conditioning temperature must be precisely matched to each other. The conditioning temperature must not be too high, as this can lead to the build-up of an excessive pressure due to the low boiling point of the water in the urine, and this can cause the vial to burst. In blood samples, a high conditioning temperature can lead to coagulation of the blood, and this can interfere with the determination of the components. However, the temperature must also not be too low, as low-volatility substances cannot then be determined reproducibly. The influence of the conditioning temperature is explained in more detail below using the example of volatile organic chlorinated compounds (VOCs) in blood. The effect of various conditioning temperatures is illustrated in Fig. 10-2. The fundamental parameters were established with the aid of animal blood.

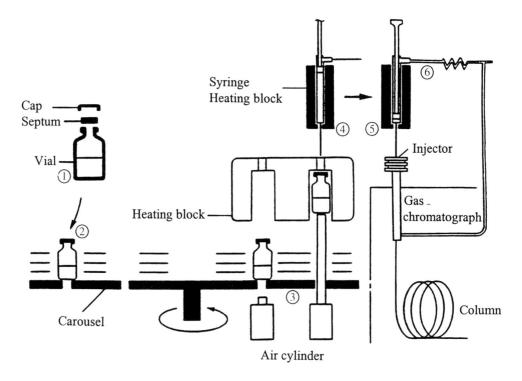

Fig. 10-1. Schematic diagram of the headspace system HSS-2B

As can be seen from Fig. 10-2, the measured peak areas are considerably higher at 60°C than at 40°C for the same conditioning time. Temperatures higher than 60°C should not be used, as these would lead to the coagulation effect mentioned above.

The influence of the conditioning time on the peak areas is illustrated in Fig. 10-3, which shows that a conditioning time of 10 min leads to peak areas that are considerably smaller than after 30 min. After a conditioning time of 10 min the system is in a reproducible state, but it is advantageous, especially for trace analysis, to use a longer time, as this causes the equilibrium to shift such that a higher proportion of the components under investigation is present in the gas phase.

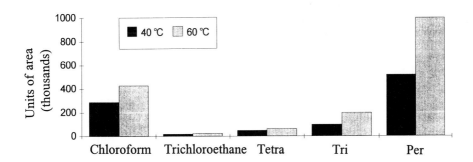

Fig. 10-2. Peak areas as a function of conditioning temperature

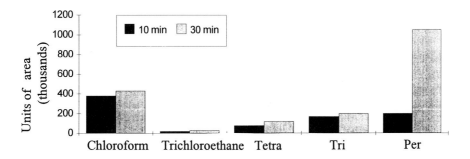

Fig. 10-3. Peak areas as a function of conditioning time

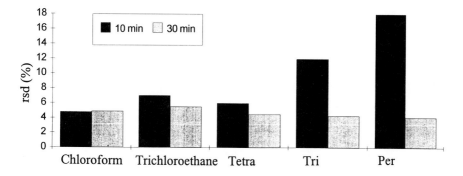

Fig. 10-4. Reproducibility of results expressed as percentage rsd (relative standard deviation).

Figure 10-4 shows the reproducibility of results using the example of VOCs in blood. Here, 2 ml blood was introduced into a headspace vial and measured under similar conditions. For each conditioning time, six samples were measured, and the relative standard deviation (rsd) was calculated from these measurements. Figure 10-4 clearly shows the difference in reproducibility after conditioning times of 10 min and 30 min at constant temperature. The system is not yet in stable equilibrium after 10 min, and this causes poor reproducibility. After 30 min a reproducible equilibrium is reached, as is clearly shown by the relative standard deviations. The average standard deviation is 3%. With standard deviations of this magnitude, the influence of the method of introduction of the sample into the headspace vial is significant. Sample introduction was by a single injection of 2 ml.

A method development of this kind should be carried out for every component and matrix to be investigated.

10.2 Equipment Parameters

The following components were used to carry out headspace analyses:

- Headspace system HSS-2B
- Gas chromatograph GC-14A
- Evaluation system C-R4AX.

All this equipment is produced by Shimadzu, Duisburg.

Table 10-1. Equipment parameters for the analyses

Chromatographic column	DB-VRX 30 m × 0.32 ID
	Film thickness 1.80 µ (manufactured by JW)
Detectors	ECD and FID
Injector	SPL-14
Carrier gas	Nitrogen 0.75 bar
Split	30 ml/min
Purge	10 ml/min
Air	0.5 bar
Hydrogen	0.5 bar
Make-up gas	1.0 bar
Headspace sample container	25 ml volume (Shimadzu)

10.3 Determination of Benzene, Toluene and Xylenes (BTX)

The components known as BTX are aromatic compounds. Benzene, which is present in gasoline and motor vehicle exhaust gases, was formerly a starting compound for the production of insecticides (pentachlorophenol), styrene, and various plastics. It was also widely used in paints, paint strippers, varnishes and adhesives. It can enter the human organism via the lungs, skin, and gastrointestinal tract. Acute intoxication leads to the following symptoms:

- Irritation of the mucous membranes
- Headache.

Chronic intoxication leads to the following clinical picture:

- Damage to the bone marrow (carcinogenic effect)
- Changes to the blood count
- Damage to the liver, spleen, kidneys and genetic material.

Benzene is metabolized to phenol in the human organism. The phenol is determined from the urine. Recent research has shown that the determination of S-phenylmercapturic acid and *trans,trans*-muconic acid in the urine is more informative than the determination of phenol.

The following limit values apply to benzene:
Exposure equivalent: 0.9 µg/l in whole blood at 1.0 mg/m^3 air
Exposure equivalent: 5 µg/l in whole blood at 3.3 mg/m^3 air
Exposure equivalent: 14 µg/l in whole blood at 6.5 mg/m^3 air
MAK: 16 mg/m^3 in the latest regulations (1971)
BAT: Not available [8]
Normal values: < 0.19 µg/l for nonsmokers
                    < 0.49 µg/l for smokers.

The chromatogram of a blood sample containing benzene is given in Fig. 10-5.

Toluene is used as a starting material in the production of aromatic chemical intermediates. It is used as a solvent for fats, paints and varnishes, and is also used in the explosives and adhesives industries. Entry into the human organism is analogous to benzene. Toluene is, moreover, an addictive substance which is used by "glue sniffers", who inhale the vapor emitted by adhesives. The clinical picture resembles that for benzene inhalation. The change in the toluene level in blood, after exposure, proceeds in several phases, which are shown in Table 10-2.

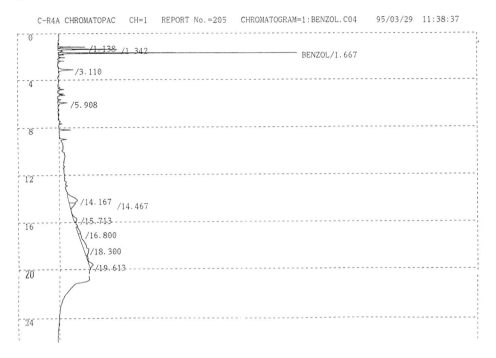

Fig. 10-5. Chromatogram of a blood sample with 25 µg/l benzene.

The following limit values apply to toluene:
BAT: 1.7 mg/l in whole blood
Normal value: < 5 µg/l
MAK: 100 ppm.

Table 10-2. Changes in levels of toluene in blood

Time after exposure	Half-life period
< 1 h	30 min
15 h	7.5 h
> 24 h	Completely eliminated

Xylene is the most important aromatic solvent used in the paint, varnish and adhesives industries, and is also a starting material in the production of plastics. It enters the human organism by the same routes as benzene and toluene, and the clinical picture is similar to that for toluene.

The following limit values apply to xylene (all isomers):
BAT: 1.5 mg/l in the blood
Normal value: < 5 µg/l
MAK: 100 ppm

Methylhippuric (toluric) acid (metabolite of xylene)
BAT: 2 g/l in urine.

Table 10-3 gives the parameters that were used in the practical analysis for BTX aromatics.

Table 10-3. Equipment parameters for the BTX aromatics

Conditioning temperature	80°C
Conditioning time	30 min
Injection volume and syringe temperature	1 ml, 90°C
Injector temperature	200°C
Temperature program	50°C, 4 min isotherm,
	Heat at 10°C/min to 200°C
Amount of sample	2 ml EDTA-blood
Detector	FID
Detector range	10^0
Attenuation	10^3
Calibration	Addition of standard or external calibration

10.4 Determination of Phenol

Phenol is the main metabolite of benzene. It occurs mainly in foamed organic products, plastics, adhesives, tar and tarred roofing felt. Phenol is also a physiological metabolic product of intestinal bacteria. Entry into the human organism is analogous to that of benzene. The clinical picture corresponds to that of the BTX aromatics.

The following limit values apply to phenol:
BAT: 300 mg/l urine
MAK: 19 mg/m^3 [9]
Normal value: < 10 mg/l

The parameters used in urine analysis for phenol are listed in Table 10-4. Figure 10-6 shows the chromatogram obtained in this investigation.

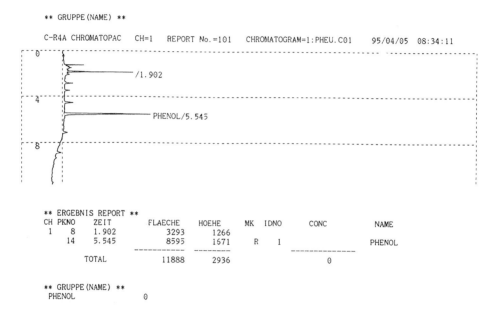

Fig. 10-6. Chromatogram of a urine sample with 27 mg/l phenol.

Table 10-4. Equipment parameters for phenol

Conditioning temperature	90°C
Conditioning time	60 min
Injection volume and syringe temperature	1 ml, 95°C
Injector temperature	200°C
Temperature program	120°C isotherm, 6 min
	Heat at 15°C/min to 200°C
Amount of sample	2 ml urine and 20 µl conc. H_2SO_4
Detector	FID 300°C
Detector range	10^0
Attenuation	10^4
Calibration	Addition of standard or external calibration

10.5   Determination of Volatile Chlorinated Organic Compounds (VOCs)

Chlorinated organic compounds are used in large quantities in the paint, varnish and chemical industries because of their remarkable properties as fat solvents and their nonflammability. Entry into the human organism is via the lungs, the skin, and the gastrointestinal tract. The clinical picture includes irritation of the mucous membranes, narcotic effects, liver damage (fatty infiltration), and damage to the heart muscle and kidneys. Also, chlorinated hydrocarbons can cause the CHC syndrome.

Chloroform and carbon tetrachloride were formerly used as solvents, but the compounds mainly used today are dichloromethane, trichloroethene, tetrachloro-ethene and 1,1,1-trichloroethane. The III B list includes a comment that the above-mentioned chlorinated hydrocarbons have carcinogenic potential.

Chloroform
This is formed during the chlorination of water and during the breakdown of other chlorinated hydrocarbons (CHCs).
No limit value for blood is currently available in the literature.
Limit of detection: 2 µg/l
MAK: 10 ppm

Dichloromethane (methylene chloride)
This is metabolized to carbon monoxide (CO), which is toxic by inhalation.
Limit of detection: 20 µg/l
BAT: 1 mg/l
BAT for CO-HB: 5%
MAK: 100 ppm

Trichloroethene
This is metabolized in the body to trichloroacetic acid, which can be determined in the urine.
BAT: 5 mg/l
MAK: 50 ppm
Limit of detection: 2 µg/l

Tetrachloroethene
This, like trichloroethene, is metabolized to trichloroacetic acid.
Limit of detection: 2 µg/l
BAT: 1 mg/l
MAK: 30 ppm

1,1,1-Trichloroethane
Limit of detection: 2 µg/l
BAT: 550 µg/l
MAK: 200 ppm

Table 10-5 gives the equipment parameters used for the practical determination of VOCs. A blank chromatogram of an untreated blood sample is shown in Fig. 10-7.

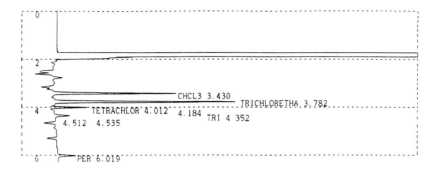

Fig. 10-7. Blank chromatogram of an untreated blood analyzed for volatile halogenated organic compounds

Table 10-5. Equipment parameters for VOCs

Conditioning temperature	60°C
Conditioning time	30 min
Injection volume and syringe temperature	1 ml, 70°C
Injector temperature	200°C
Temperature program	50°C isotherm, 4 min
	Heat at 10°C/min to 200°C
Amount of sample	2 ml EDTA-blood and 0.5 ml water
Detector	ECD 300°C
Detector range	10^0
Attenuation	10^4
Calibration	Addition of standard

10.6 Determination of Trichloroacetic Acid (TCA)

On heating, trichloroacetic acid decomposes to chloroform and carbon dioxide. Analysis is by determination of the chloroform formed.

Limit of detection:	1.0 mg/l
Normal value:	< 60 mg/l
BAT:	100 mg/l

Table 10-6 gives an overview of the equipment parameters used.

Table 10-6. Equipment parameters for trichloroacetic acid

Conditioning temperature	90°C
Conditioning time	3 h (can alternatively be carried out in the water bath under the same conditions)
Injection volume and syringe temperature	1 ml, 95°C
Injector temperature	200°C
Temperature program	50°C isotherm, 6 min
	Heat at 10°C/min to 200°C
Amount of sample	2 ml urine
Detector	ECD 300°C
Detector range	10^0
Attenuation	10^8
Calibration	External calibration and checking with standard urine from the company Biorad

10.7 Determination of Blood Alcohols

Alcohols are used as organic solvents, cleaning agents, and starting materials for the production of esters, ethanol being also a recreational beverage.

Methanol occurs in high concentrations in brandy and other drinks (distillates from fermented fruit). It is metabolized to formaldehyde and then formic acid, which, because it is eliminated slowly, can lead to severe acidosis (acidic products of metabolism).

Limit of detection: 1 mg/l
MAK: 200 ppm
BAT: 30 mg/l (urine)

Ethanol is determined to assess ability to drive a vehicle, and it must also be
monitored in patients undergoing methadone substitution therapy or who are
addicted to medicaments, as alcohol can intensify the effects of such
medicaments. Table 10-7 lists the parameters used in the analysis, and Fig. 10-8
shows a chromatogram obtained using these conditions.

Table 10-7. Equipment parameters for ethanol

Conditioning temperature	80°C
Conditioning time	30 min
Injection volume and syringe temperature	1 ml, 90°C
Injector temperature	200°C
Temperature program	40°C, 5 min isotherm, heat at 10°C/min to 200°C
Detector	FID 300°C
Detector range	10^0
Attenuation	For ethanol 10^7, for methanol 10^0
Amount of sample (ethanol)	500 µl serum or EDTA-plasma and 500 µl water
Amount of sample (methanol)	1 ml serum or EDTA-plasma and 1 ml water
Calibration	Addition of standard or external calibration

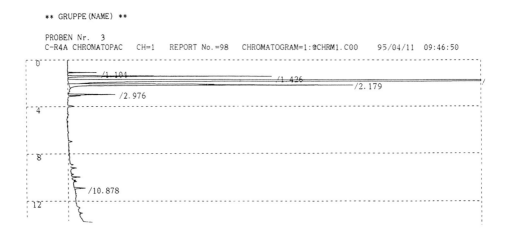

Fig. 10-8. Chromatogram of a serum sample with 1.7% ethanol.

References for Part III

[1] Hauptverband der gewerblichen Berufsgenossenschaften, BIA-Report, Gefahrstoff-listen, 1996.

[2] Parlar, H., Gebeflügi, I., Verhalten und Vorkommen von PCP in geschlossenen Räumen, In: Aurand, K., Hässelbarth, U., Lahmann, E., Müller, G., Niemitz, W., *Organische Verunreinigungen in der Umwelt*, Berlin: M.E. Schmidt Verlag, 1978.

[3] Brunner, H., Diss. Eberhard-Karls-Universität Tübingen, 1990.

[4] Wichmann, H.E., Schlipköter, H.W., Fülgraff, G.M., *Handbuch der Umweltmedizin*, Landsberg/Lech: Ecomed, 1995.

[5] Daunderer, M., *Umweltgifte*, Landsberg/Lech: Ecomed, 1995.

[6] Pflanzenschutz-Höchstmengen-Verordnung vom 24. 6. 1982.

[7] Pahlmann, W., Bioscientia, Abt. Pharmakologie, Toxikologie und Umweltanalytik, Pentachlorphenol im Blutserum, Methode 1.0/1989.

[8] BIA-Report, Gefahrstoffliste 1996, *Gefahrstoffe am Arbeitsplatz*, Berlin: Erich Schmidt Verlag, 1996.

Part IV   Organization and Cost-Effectiveness
in the Clinical Laboratory

11. The Provision of Samples to the Medical Laboratory

A. Wertmann

The provision of samples to the medical laboratory starts with the taking of a sample of the material under investigation by the doctor or the patient and finishes with the handing over of the sample to the analyst. It is an important part of the preanalytical phase of every laboratory investigation, and is one of the most decisive factors contributing to the accuracy of the eventual measurement result [1,2]. The provision of the sample thus includes:

1. Taking the sample

- Selecting the material to be examined
- Establishing the time at which the sample is taken
- Selecting the sample container

2. Transport and storage of the sample

- Intermediate storage until the sample is transported
- Maintenance of the correct transport conditions
- Storage until analysis is performed.

Sample Materials

The medical laboratory deals mainly with blood and urine samples. These two matrices are described in more detail later. Other sample materials can include:

- Gastric juice
- Mother's milk
- Samples obtained by hypodermic syringe
- Feces
- Others

Apart from the methods of obtaining these materials, the points to bear in mind with respect to their shipment to the laboratory are very similar for each material.

11.1 Timing

Not only are the technical aspects of sample taking important, but so also is the timing of this operation in order to maximize the usefulness of the analytical results, and, in fact, strict rules (e.g. use of the fasting value) for analyses performed in the course of investigations into the metabolism of fats and carbohydrates have been routinely followed for many years.

Similar considerations must also apply to the time at which samples are taken for the determination of pollutants, in therapy control, and in the detection of abuse of drugs and medicaments. For example, many hydrocarbons can be determined in blood only if samples are taken during or shortly after exposure, as exhalation, metabolism, excretion, and deposition in body tissues can quickly cause the blood concentration to fall below the limit of detection.

Also, drugs and medicaments can break down rapidly in the body, depending on their chemical structure. This leads to large changes in the blood concentration of many substances, depending on the time at which the sample is taken.

11.2 Sample Taking

The method of taking samples for analysis in the fields of toxicology and industrial and environmental medicine is of great importance. Many important analytes in these fields are present in only very small traces in the material being analyzed.

Loss of analyte or increase in its concentration due to its chemical or physical properties must be prevented. The full benefits of advances in the sensitivity and precision of the analytical techniques used in these fields can only be effectively realized if the taking of the sample and its transportation and storage are also performed as well as possible.

11.3 Blood Samples

A blood sample is taken from the patient by the doctor or by some other authorized person. This is to guard against manipulation of the sample by the patient or contamination from the place where the sample is taken.

Also, the analysis from blood enables a conclusion to be drawn concerning the concentration of an analyte and the possible effect on the concentration of the time at which the blood sample is taken. The correct labeling of both the sample container and the package that contains it and provision of the necessary documentation for the laboratory ensure that the sample will be processed without delay.

A blood sample can be taken with the aid of a disposable cannula or syringe, and various other systems are available. The so-called closed system gives good protection from contamination after the sample has been taken, as it is not necessary to transfer it into another vial for shipment. Nevertheless, a blank value should be determined for each analysis. This is also necessary for sample taking with a cannula or a syringe, especially if the sample tubes contain additives (serum separating agents, anticoagulants, stabilizers).

Sample vials can be made of various materials, depending on the supplier of the system for obtaining the sample. This can lead to problems if, for example, they have to be made of glass or other special materials to prevent loss of a particular analyte and they are not availkable from the manufacturer. Unfortunately, the various blood sampling systems are usually incompatible, so that the special vials are effectively non-interchangeable.

This and other questions relating to the taking, storage or transport of blood samples should preferably be discussed with the laboratory, which should be able to make available any sampling apparatus and transport containers necessary for special analyses. These are then tested for suitability for the analyses concerned. The conditions for taking the sample, possible methods of intermediate storage until shipment, and transport conditions should thus be mutually compatible.

11.3.1 Choice of Sample Container

For most blood analyses, the commercially available plastic vials and transport containers are suitable for transportation, with addition of appropriate additives as required. The following sample materials require special containers:

- Solvents
- Hydrocarbons
- Halogenated hydrocarbons
- Alcohols
- Polychlorinated biphenyls (PCBs)
- Wood preservatives
- Heavy metals
- Trace elements.

For the accurate determination of these analytes, the samples must be transported in special glass vials with addition of suitable anticoagulants, as these substances can be adsorbed by plastic materials (cap or stopper coated with Teflon®). Such sample tubes can be supplied by the laboratories on demand.

Exogenic contamination of the sample by the steel cannula, serum separating agent or other additives or the material of the sample tube is possible.

Contamination with metals, metal ions and other substances is also observed with containers of this type if they are not supplied by the manufacturer with the closures already fitted to the sample tubes (i.e. if the tubes and closures are packed separately). Under some circumstances, contamination of the substance to be analyzed by dust, aerosols and vapors can occur on long storage.

The danger of contamination effects in determinations of aluminum in blood deserves special mention. Some manufacturers of blood sampling apparatus today supply special sample vials for the determination of metals in blood. Tests show that these lead to very little contamination of the sample by the vials. (These too can be obtained from the laboratories on special request.)

Glass tubes are unsuitable as vials for the containment and transportation of samples for the analysis of metals in blood and other materials, as some ions migrate from the walls of the vial into the sample material on long storage.

Many parameters in the medical laboratory are determined from serum or plasma. This should be obtained by removing the blood coagula and corpuscles as soon as possible after the blood sample has been taken. This prevents any analytes from migrating from the solid constituents of the blood into the serum or plasma (e.g. potassium and some enzymes), prevents migration in the reverse direction (e.g. of chloride), and also prevents chemical reactions in vitro from causing increase or decrease of the concentrations of any substances [3].

11.3.2 Urine Samples

Urine samples are collected in new plastic vessels (usually of polyethylene or polycarbonate). These can have a volume of 50–2000 ml, as required. If these containers are also used for transport of the sample, it is essential that they should have a screw closure with a seal. Containers with snap closures or similar are unsuitable for transportation. Even small mechanical stresses or temperature changes can affect the integrity of the seal (with possible escape of potentially infectious material or contamination of the sample). The sample should be marked with the name of the donor and the sampling time or period.

For the determination of the consumption of drugs and medicaments, urine is more suitable than blood as the material for analysis, as the substances of interest and their metabolites can be determined in urine even after several days, whereas they can often break down in blood within a few hours.

In the determination of trace elements and heavy metals from urine, the collection vessel should previously be washed with dilute nitric acid followed by distilled water.

Urine samples are often obtained in a domestic environment (24-h collection period) or, in the case of industrial medicinal investigations, in a working environment, so that location-dependent exogenic contamination can occur.

11.4 Sample Transportation and Storage

To prevent any changes to the sample matrix due to storage and transportation, it is important to let the laboratory have the sample immediately after it has been taken, so that the analytical work can be started. As this is not possible in most cases, the following recommendations should be observed for short-term storage before transportation or storage until analysis:

- Thorough mixing of the sample with the added anticoagulants or stabilizers
- Removal of serum or plasma if this is appropriate for the analysis required
- Storage of urine in suitable closed containers in a refrigerator at between +2 and +4°C (also during collection periods!)
- Storage of bood, serum, plasma or urine for longer periods at −20°C in suitable closed containers if such storage is acceptable for the analysis required.

11.4.1 Transport Conditions

If samples are to be delivered by post, postal regulations for medical and biological experimental materials must be observed [4]. For liquid experimental materials (e.g. blood, serum, urine, feces), these are as follows:

- The sender must ensure that the consignment is packed in such a way that it reaches the analytical laboratory in good condition and in no way constitutes a danger for humans, animals or the environment.
- Postal deliveries of blood sampling systems must not include the injection needle.
- Postal packages must comply with DIN 55 515 Part 1.

This DIN Standard includes the safety specifications for the packaging of medical samples, according to which a package for delivery by post must consist of:

- The inner packaging (sample vials containing the material for analysis)
- The absorbent material (to absorb the contents of the sample vial in case of breakage)
- The outer casing (which gives protection against mechanical stress and contains the inner packaging and absorbent materials)
- The outer wrapping (which must comply with the regulations for postal delivery of medicinal experimental materials and contains the outer casing).

All the necessary materials of the prescribed quality are supplied by manufacturers of medical requisites or by medical laboratories.

The sender has legal responsibility for the delivery [4].

If possible, postal packages should not be consigned at the weekend. Many laboratories use messenger and courier services which guarantee transport of the sample to the laboratory in the shortest possible time. The sender is recommended to use this method for urgent analyses, working in coordination with the laboratory, in order to minimize harmful effects during sample deliveries. Properly prepared packages, as for postal deliveries, should be used.

12. Internal Cost Calculations in GC/MS Analysis using Drug Screening from Urine as an Example

Jürgen Sawazki

The total costs of a drug screening by GC/MS can be subdivided into the following cost groups:

1. Machine costs
2. Material costs of sample preparation
3. Personnel costs
4. Space costs
5. Quality assurance costs.

These cost groups are described below. By considering the various types of cost in GC/MS analysis, it is easily possible, for example, to recognize synergistic effects within an existing cost structure. It should be mentioned here that material and personnel costs can vary over a wide range. Average prices should be used in computerized studies of nonpersonnel costs. The parameter "useful life" should be estimated conservatively in order to give a realistic rather than a very optimistic costing. Personnel costs are set slightly above the average so as to make provision for the cost of well-qualified staff.

The cost calculation shown here can be applied to any particular laboratory by substitution of the locally applicable cost figures, and should give a realistic picture if the fundamental basis of the calculation is borne in mind. In this situation, the actual extent of utilization of personnel should be checked before coming to any conclusion about the profitability of the GC/MS analysis.

12.1 Machine Costs per Analysis in GC/MS Analysis

Machine costs include the proportional costs of procurement and all operating costs incurred by the use of the equipment.

An estimate is often made of the machine costs per sample invoiced. The possible use of the competitively priced quadrupole mass spectrometer rather than a sector field machine was not considered in these examples. The present

Chapter should help to dispose of the misconception that this method of analysis is expensive.

The proportional machine costs per sample tested depend directly on the number of analyses per year.

12.1.1 Principles of the Calculations

12.1.1.1 Proportional Procurement Costs

The calculation of internal costs is based on the estimated total useful life of the equipment for determination of the proportional equipment costs per year not including the tax benefits of depreciation. For an expected total useful life of 10 years, a purchase price of 100 000 DM gives proportional machine costs of 10 000 DM/year. It is assumed here that the extent of utilization of the machine remains the same during the whole of its useful life (see also Table 12-1).

Table 12-1. Method of calculation of machine costs of GC/MS

Procurement costs		
Purchase price	100 000.00 DM	
Estimated useful life	10 years	
Annual depreciation		10 000.00 DM
Operating costs		
Carrier gas: Helium 5.0, 50 l, 200 bar	1 year	600.00 DM
Column: e.g. 25 m SE-54-CB, 0.25 mm I.D., 0.25 μm film	800 measurements	650.00 DM
Washer plus ferrule (interface)	800 measurements	52.00 DM
Thermogreen septa	100 injections	8.30 DM
Silanized insert plus graphite ferrule	7 days	31.50 DM
Filaments	2500 injections	500.00 DM
Electricity for 20 measurements/day, 0.25 DM/kWh Oven temperature 50–320°C	1 month	400.00 DM
Maintenance		
Change of pump oil Cleaning of injector, split, quadrupole and prefilter, pump pipework	1 year	2 000.00 DM

12.1.1.2 Operating Costs

Carrier Gas

In the calculation of the gas consumption, it is assumed that this is regulated by a modern controller of the type installed in the latest gas chromatographs. The programmable control of the split reduces gas consumption by the equipment and hence considerably reduces the cost of the carrier gas (helium).

Chromatographic Column

The useful life of the column is subject to many influences. When analyzing prepared urine samples for addictive substances, we make 800 measurements using a 25 m SE-54-CB column with an inside diameter of 0.25 mm and a film thickness of 0.25 μm.

Many users shorten the column by a small amount at each end after 600–800 measurements, as the film quality usually deteriorates initially at the interface with the mass spectrometer and close to the injector. This procedure reduces the costs of the washer plus ferrule. No figure for the added useful life of a shortened column can be stated, as we have no experience of this. These potential cost reductions are therefore not included in the calculation.

Thermogreen Septa

Every time a measurement is made, the septum of the injector is pierced so that the sample can be injected. However, the septum cannot repeatedly reseal itself perfectly for an indefinite period of time, and it is therefore recommended that it should be changed after 100 injections to prevent air from passing into the injector.

Glass Insert

To increase the sensitivity of measurement of many substances (e.g. morphine, oxazepam, amphetamine), it is strongly recommended that the silanized glass insert and the silanized glass wool which it contains should be changed every week. Inserts and ferrules can be used many times, but prices used in the calculation are of new items, labor costs of cleaning and silanizing used inserts for reuse being accounted for as expediency costs.

Filaments

In the analysis of addictive substances from urine, it is assumed that the filament is exchanged after 2500 measurements, as it is possible for samples to contain very high levels of drugs or accompanying substances.

Energy Costs

An above-average electricity consumption (1600 kWh/month) has been assumed to allow for the treatment of large amounts of sample or long heating phases. A cost of 0.25 DM/kWh is assumed.

Maintenance

Our maintenance system includes a change of pump oil and cleaning of the injector, the split, and the connection between the mass spectrometer and the pump at 6-monthly intervals, while the quadrupole and the prefilter are cleaned once a year only. This maintenance work is carried out by the customer services department.

We clean the ion source and lens system ourselves as necessary (approximately every 3 months).

Repair Costs

These costs can only be approximated for the total lifetime of the equipment, as they vary greatly according to the quality of the equipment and the qualification of the operator. To allow for the fact that the more heavily the equipment is used the more likely it is to require repair, a repair cost of 1.00 DM per measurement is included in the calculation.

12.1.2 Determination of Machine Costs per Measurement Result

In this calculation, three cost blocks can be constructed (see Table 12-2).

Table 12-2. Calculation of machine costs per measurement result

Example: 5000 measurements/year		Costs per year
A) Purchase price of the equipment Estimated useful life	(100 000 DM) (10 years)	10 000.00 DM
B) Carrier gas		600.00 DM
Electricity		4 800.00 DM
Maintenance		2 000.00 DM
Silanized inserts		1 638.00 DM
C) Column including ferrule	(1 column per 800 measurements)	4 387.50 DM
Septa	(Change after 100 injections)	415.00 DM
Filaments	(1 filament per 2 500 measurements)	1 000.00 DM
Repair costs	(1.00 DM per measurement)	5 000.00 DM
	Total costs per year:	29 840.50 DM
	Costs per measurement:	5.97 DM

12.1.2.1 Proportional Procurement Costs

When the useful life of the equipment has been established, the proportional procurement costs per year can be found. If the equipment is used at a constant rate over the course of its useful life, these are independent of the number of measurements. The determining factors are simply the purchase price of the equipment and its estimated useful life.

12.1.2.2 Carrier Gas, Electricity, Inserts and Maintenance

A GC/MS unit reaches its maximum measurement sensitivity many hours after start-up. Because of this long equilibration time, the system is normally kept continually ready for operation, and costs for carrier gas and electricity are therefore incurred during this period also. However, the program-controlled decrease in the carrier gas flow rate and the use of suitable temperature programs enable the costs for the readiness period, when no measurements are being made, to be lower than those during the measuring period. The need to replace the silanized insert is also fairly independent of the number of measurements, as it is assumed as a basis of the calculation that the periods between replacements are fixed. These periods need only be reduced if samples for analysis are being received at a very high rate. Intervals between routine maintenance operations by the customer service department are usually independent of the number of measurements. Only if samples for analysis are arriving at extremely high rates or if they contain very high levels of analytes are shorter intervals between maintenance operations likely. However, these computerized calculations may be simplified by neglecting the small variations in the costs of carrier gas, electricity, silanized inserts and maintenance.

12.1.2.3 Columns, Septa, Filaments, and Unplanned Repair Work

The costs of chromatographic columns and septa depend directly on the number of measurements. The costs of unplanned repairs depend on the age of the equipment and its extent of utilization.

If the proportional costs of procurement are added to the costs of the carrier gas, electricity, inserts and maintenance, and if the GC/MS is always kept switched on and ready for use, the fixed costs per year are ca. 19 000 DM, using the example of the calculation given above. This represents very low variable costs for columns, septa, filaments and repairs, namely 2.16 DM for each measurement carried out.

The machine cost per measurement as a function of the extent of use of the equipment is shown in Table 12-3. In this calculation, it is assumed that the machine is operated on 250 days of the year. The calculations are as shown in Table 12-2. A medium work rate of 1000–5000 measurements per year was assumed. For a very high or a very low extent of utilization of the GC/MS, a modification of the fundamental basis of the calculation is necessary. As can be seen from Table 12-3, the machine costs per measurement result are 21.20 DM if there are only four measurements per working day. This amount decreases considerably with a small increase in the equipment work rate, becoming < 7.00 DM per measurement result for 16 measurements per working day. A graph of the GC/MS machine costs of a measurement result plotted against the number of urine samples analyzed per year is shown in Fig. 12-1.

12.1.3 Limits of the Capacity of a GC/MS

If the equipment is provided with an automatic sample injector (Autosampler), the use of the method described in Chapter 9 for drug screening enables considerably more than the 20 measurements per working day shown in Table 12-3.

One analysis run takes 20 min, and the system is ready to measure the next sample after a total elapsed time of 30 min at the most. This means that the maximum capacity per day is 48 measurements, corresponding to a capacity limit of ca. 40 samples. The other 8 analytical runs are required for the quality assurance of the analysis. Based on 250 working days, this method shows that 10 000 drug screening measurements should be possible per year.

On the other hand, these calculations indicate that the treatment of more than 10 000 samples per year per machine is unrealistic, as periods during which the equipment is out of commission for the exchange of columns, inserts and septa and for all the many and various types of maintenance work are unavoidable. Continuing to use the above example, if more than 10 000 measurements per year are required in the field of drug screening, the calculation shows that the procurement of a second GC/MS must be considered.

Table 12-3. Effect of use intensity on machine costs

Measurements per working day	Measurements per year	Total machine costs per year	Costs per measurement
4	1 000	20 998.50 DM	21.20 DM
6	1 500	21 978.75 DM	14.21 DM
8	2 000	22 959.00 DM	11.68 DM
10	2 500	23 939.25 DM	9.78 DM
12	3 000	24 919.50 DM	8.51 DM
14	3 500	25 899.75 DM	7.60 DM
16	4 000	26 880.00 DM	6.92 DM
18	4 500	27 860.25 DM	6.39 DM
20	5 000	28 840.50 DM	5.97 DM

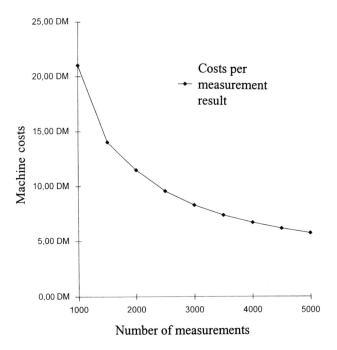

Fig. 12-1. Machine costs of GC/MS per measurement result.

12.1.4 Conclusions

Maintaining the equipment in a permanent state of readiness incurs considerable fixed costs, but increasing the extent of utilization of the GC/MS leads to disproportionately high cost degression effects with respect to machine costs per measurement result, particularly if the number of samples analyzed per year is low. The fact that the equipment can be put to a wide range of uses can therefore lead to lower machine costs per measurement result.

The very low variable costs, 2.16 DM per measurement, make it unlikely that additional measurements (e.g. SIM runs) to provide more comprehensive results for an investigation would be ruled out for economic reasons. The very low variable costs of these analyses are also very advantageous when carrying out the necessary quality assurance runs.

12.2 Material Costs of Sample Preparation

The method described in Chapter 7 is used to calculate the cost of the sample preparation. The relevant costs are as follows:

1 Toxi-Tube A (DRG Instruments)	5.68 DM
Methanol, acetic anhydride	0.28 DM
Articles used once (vial, glass, etc.)	0.90 DM
	6.86 DM

Solid phase extraction with Bond-Eluat-Certify® cartridges is rather less expensive than extraction with Toxi-Tubes®, but is somewhat more time-consuming. The use of "Eppendorf" vessels instead of the Autosampler vials gives a further saving of ca. 0.50 DM per sample. We have shown in our laboratory that 5 ml glass containers can be used to produce cheap ampoules that are quite suitable for vaporization of the organic phase.

12.3 Personnel Costs

In the calculation of personnel costs, it is assumed that sample pretratment is performed by less highly qualified staff. The hourly rate for these workers is 40.00 DM including incidental labor costs. For preparation of GC/MS analysis up to the point of injection, a working time of 10 min/sample gives a cost of 6.66 DM.

The evaluation of the analysis is performed by highly qualified staff earning 80.00 DM/h inclusive of supplementary costs. An experienced analyst will evaluate a GC/MS run in an average time of 10 min. The cost of this is 13.32 DM.

12.4 Cost of Laboratory Space

The amount of additional laboratory space that should be set aside for installation of GC/MS analysis equipment is small. Therefore, in this calculation, only the marginal cost of allocating extra space in an existing laboratory is considered, and not the cost of providing new laboratory accommodation.

The sample preparation should be carried out under a fume hood, which is usually already available in the laboratory.

Installation of a GC/MS requires an area of 150×80 cm of laboratory bench space, and it is therefore essential that the location should allow for the heat evolved by the gas chromatograph. For the evaluation of the chromatograms, a writing desk with space for a computer and printer should be provided. The equipment should be accessible from all sides for maintenance purposes. If no central gas supply exists, space for a helium bottle is also required. A total floor area of approximately 8 m^2 should be sufficient for the installation of a GC/MS.

The exact costs of the laboratory space can only be determined for each individual case. In this calculation, a covering cost of 1.00 DM per analysis should be included to allow for miscellaneous labor costs and reception of the GC/MS samples.

12.5 Total Costs

All the costs of a drug screening by GC/MS are listed in Table 12-4. To determine the cost per measurement, it is assumed that 5000 patient samples are tested per year.

Table 12-4. Total costs of GC/MS analysis per measurement

1. Machine costs (for 5000 measurements per year)	5.97 DM
2. Material costs of sample preparation	6.86 DM
3. Personnel costs (including sample preparation)	19.98 DM
4. Costs of laboratory space	1.00 DM
5. Quality assurance costs	3.38 DM
Total costs:	**37.19 DM**

Immediately after installation of GC/MS drug screening, the number of samples tested will be lower than that shown in the example calculation. The machine costs (and hence the total costs) per measurement will therefore tend to be higher, although this may be compensated for by a synergy effect in personnel costs, as the existing personnel will be able to do the additional work if the number of samples received is small. When determining the costs of a chargeable sample, it must not be forgotten that additional nonchargeable measurements must be carried out for quality assurance. In our laboratory, the number of measurements made for quality assurance reasons amounts to one fifth of the number of measurements made on patient samples. As not all measurements for quality assurance require a sample preparation and the evaluation of the measurement of a standard sample often requires less time than the identification of all the components of a patient sample, the costs of quality assurance are allowed for by adding 10% of the total costs per measurement result estimated up to this point. The total costs per chargeable patient sample thus amount to 37.19 DM in the example calculation.

12.6 Discussion of the Results

The economics of GC/MS analysis depends on the analysis required and on the viewpoint of the observer. Thus, for example, GC/MS is not a viable method for determining cannabis from urine in drug screening, as a separate sample treatment is necessary here, and a request for analysis can usually be dealt with adequately, from both economic and quality aspects, by an immunological test. This does not apply to a comprehensive screening for addictive substances. In this case, immunological tests for cannabis can only give information about the following addictive substances or groups of substances: amphetamines, barbiturates, benzodiazepines, cocaine, methadone, methaqualone, opioids and phencyclidine. The total costs of the immunological test are at least as high as those of a GC/MS analysis for the same purpose. The immunological tests for amphetamines, barbiturates, benzodiazepines and opioids unfortunately only give information about groups of substances. GC/MS analysis clearly gives results with a much higher information content, as single substances and/or their metabolites are always determined, and other addictive substances such as antihistamines, chlormethiazole, pethidine, tilidine or tramadol are also determined. GC/MS analyses also give information about accompanying medication. The considerably higher information content of GC/MS measurements commands a higher rate of remuneration. Thus, the fee for an analysis by GC/MS as described above, in accordance with the Einheitlicher

Bewertungsmassstab der Krankenkassen (unified remuneration scale of the health insurance schemes, EBM) has been 1300 points since 1 April 1994, while a corresponding analysis using immunological tests would earn a maximum of 1050 points.

The capabilities of GC/MS analysis bring additional benefits in our laboratory. The number of screenings has increased considerably because of the availability of this efficient analytical method, and reliable answers to clinically relevant questions can be given within a short space of time. This was at one time difficult or impossible.

13. Quality Assurance in the Clinical Laboratory using Drug Analysis as the Example

Petra Gerhards, Jörg Szigan

This chapter deals with some methods of quality assurance, an important aspect of GLP (good laboratory practice) used to monitor drug analysis in the clinical laboratory. To ensure that the analytical results are of good quality, it is essential to find suitable methods to use for the routine procedures, and, in this connection, we discuss the use of internal and external standards and propose some methods for the documentation and verification of the analytical results, the quality of the results of GC/MS analysis being at the forefront of our discussion.

It is important to maintain a laboratory journal for the GC/MS instrument. All tuning files should be documented here, along with records of the installations of columns, replacements of inserts and septa, and any other changes to the system. The analyses themselves must, of course, be documented. This provides trans-parency for other operators and for the current state of the system.

13.1 Internal Quality Assurance

Internal quality assurance includes the following:

- Blank runs
- Checking of the retention times
- Changes to the system, e.g., replacement of the chromatographic column
- Documentation of the method files used
- Tuning reports
- Documentation of the sample preparation
- Checking of the system using methanolic standards and standard urines.

The daily checking should include leak testing of the equipment and the use of an internal standard.

Once per day, the molecular masses that could indicate a leak should be checked. These are 18 (water), 28 (nitrogen) and 32 (oxygen). A high level of these masses indicates a leak in the system, e.g., at the interface between the GC and the MS. If a leak is not found by routine testing, it can lead to an increased background and consequently poor analytical results. In the worst case, it can result in destruction of the filament.

13.1.1 Standards

Before starting any analyses, it is, of course, essential to perform a blank run in order to condition the column, and, if drug analyses are to be carried out, a methanolic drug standard or a prepared (derivatized) standard urine should also be measured to check the sensitivity of the equipment and the separating power of the column. Figure 13-1 shows a standard that can be used in the clinical laboratory. The substances present in the standard should correspond to the type of analyses required in the laboratory. However, they should cover the whole range of possible retention times.

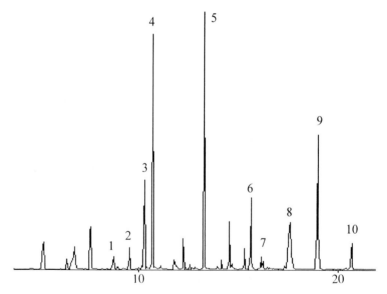

Fig. 13-1. Example of a gas chromatogram of a methanolic standard.
1 = aprobarbital; 2 = phenobarbital; 3 = caffeine; 4 = lidocaine;
5 = amitriptyline; 6 = diazepam; 7 = nordiazepam; 8 = bromazepam;
9 = flurazepam; 10 = alprazolam

The standard used here contains components representing various groups of drugs of abuse, e.g.:

- Barbiturates
- Tricyclic antidepressants
- Benzodiazepines.

13.1.2 Standard Urines

The preparation of a standard is not always easy, and standard urine samples are therefore prepared and supplied commercially. These standard urines have the advantage that they provide a check of the sample preparation procedure, while supporting the accuracy and precision of the analytical determinations. They thus enable the quality of the analyses performed in the laboratory to be assessed in the most effective manner possible.

Standard urine samples of various compositions can be obtained in various concentrations.

Suppliers of such samples include BIO-RAD and DRG Instruments. Three examples of standard urines are described below.

I. Urine Toxicology Screen Control Lyphocheck® BIO-RAD

The Lyphocheck® urine has been optimized such that the drug concentrations lie above the recommended "cut-offs" for screening determinations recommended by the Substance Abuse and Mental Health Services Administration (formerly the National Institute of Drug Abuse, NIDA) and the United States Nuclear Regulatory Commission (NRC). This standard urine is prepared from human urine to which no preservative is added and is lyophilized to improve its stability.

Table 13-1 shows the concentrations of the various components present in the urine [5].

Figure 13-2 shows the total ion chromatogram for this urine. The urine was prepared by the alkaline process and then acetylated with acetic anhydride.

Sample preparation with methanol in an acid medium causes partial methylation of benzoylecgonine, which is determined as cocaine. The morphine is acetylated by the acetic anhydride to give heroin. If alkaline sample preparation is used, it is not possible to determine THC and secobarbital. Separate sample preparations would be necessary to enable these components to be analyzed.

Table 13-1. Substances present in standard urine Toxicology Screen Lyphocheck®
             BIO-RAD (acetylated)

Analyte	Target figure [ng/ml]
Benzoylecgonine	360
Methadone	360
D-Methamphetamine	1200
Methaqualone	360
Morphine (free)	360
Oxazepam	360
Phencyclidine (PCP)	30
Propoxyphene	360
Secobarbital	360
Tetrahydrocannabinol (THC)	60

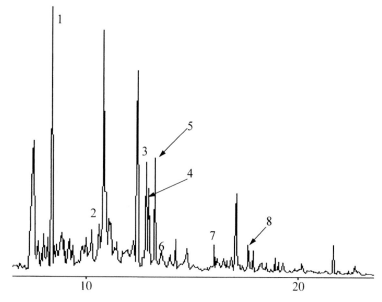

Fig. 13-2. Total ion chromatogram of the standard urine Toxicology Screen
            Lyphocheck® BIO-RAD (acetylated).
            1 = methamphetamine AC; 2 = PCP; 3 = methadone; 4 = methaqualone;
            5 = propoxyphene; 6 = cocaine; 7 = oxazepam; 8 = diacetylmorphine

II. Urine Toxicology Confirm Control Lyphocheck® BIO-RAD

This is an accuracy control standard prepared from human urine. The added drugs are present in the concentrations near to the cut-off recommended by the NIDA for confirmatory tests. The urine is in the lyophilized form. The concentrations of the components are listed in Table 13-2 [7].

Table 13-2. Substances present in standard urine Toxicology Confirm Control Lyphocheck® BIO-RAD (acetylated)

Analyte	Target figure [ng/ml]
D-Amphetamine	663
Benzoylecgonine	190
Codeine	379
D-Methamphetamine	662
Morphine (morphine-β-D-glucuronide)	356
Phenylcyclidine (PCP)	37
Tetrahydrocannabinol (THC)	18

Figure 13-3 shows the chromatogram of the non-acetylated standard urine. THC could not be determined, as a special sample preparation is necessary for its determination. Benzoylecgonine can only be determined after derivatization.

III. Standard Urine Toxi-Control No. 19, DRG Instruments

This is a qualitative standard urine. Sodium azide is added to improve its storage properties. Table 13-3 lists the concentrations of the components present [8].

Figure 13-4 shows the analysis of the standard urine No. 19. After methylation, benzoylecgonine would be determined with great sensitivity as cocaine. As well as benzoylecgonine, methylecgonine is excreted in the urine by cocaine users, and this can be determined without derivatization along with the cocaine artifact. If morphine is not detected in its underivatized form, this means that the column is showing effects of aging. Morphine is a good criterion of this, because it is polar and is adsorbed on the column if this is no longer in perfect condition.

Table 13-3. Substances present in standard urine Toxi-Control No. 19,
DRG Instruments

Analyte	Target figure [ng/ml]
Morphine	3
Amphetamine	3
Imipramine	1.5
Methadone	1.5
Propoxyphene	4
Phenobarbital	5
Secobarbital	1
Benzoylecgonine	3

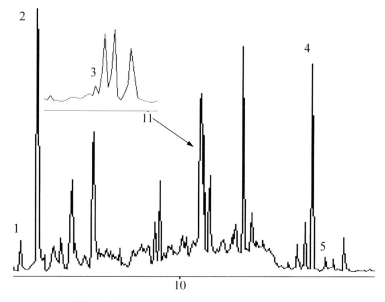

Fig. 13-3. Total ion chromatogram of the standard urine Toxicology Confirm Control
Lyphocheck® BIO-RAD.
1 = amphetamine; 2 = methamphetamine; 3 = PCP; 4 = codeine;
5 = morphine

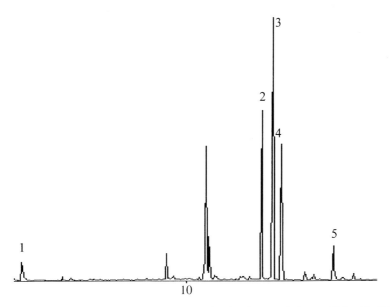

Fig. 13-4. Total ion chromatogram of the standard urine Toxi-Control No. 19, DRG
    Instruments.
    1 = amphetamine; 2 = methadone; 3 = propoxyphene; 4 = imipramine;
    5 = morphine

13.1.3 Internal Standard

A further method of checking the system and the sample preparation procedure
is by the use of an internal standard. For this, a substance must be found that is
not present in the material to be analyzed. This substance can vary from country
to country, as different drugs and drug substituents are abused in different
countries. We have found nalorphine, an opiate antagonist, to be the most
suitable substance.

Figure 13-5 shows a drug-containing urine to which nalorphine has been
added in a concentration of 200 ng/ml.

Nalorphine contains two hydroxyls, which are functional groups that can be
used to check acetylation. If the surface areas or heights of the peaks for the
internal standard change rapidly, it is probable that an error has been made in
the sample preparation. This procedure at the same time gives a check of the
statistical repeatability of the sample preparation. By plotting the peak areas in a
control chart, the user can document the reproducibility of the method. A

control chart for nalorphine is shown in Fig. 13-6. This chart also shows how many samples have been analyzed on the column.

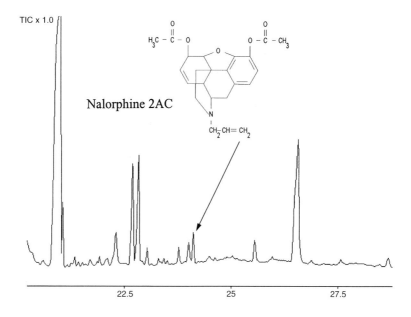

Fig. 13-5. Drug-containing urine with nalorphine (acetylated) as the internal standard

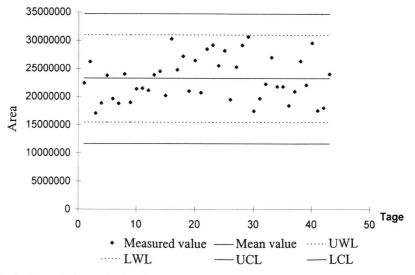

Fig. 13-6. Control chart of nalorphine (acetylated).
UWL = upper warning limit; LWL = lower warning limit; UCL = upper control limit; LCL = lower control limit

13.2 External Quality Assurance

In external quality assurance, the results of clinical laboratories are compared with each other in a procedure known as interlaboratory testing. Different types of interlaboratory tests are used for different types of analysis. Both qualitative and quantitative interlaboratory testing is used. Participation can be limited to one country or can be international.

An interlaboratory test carried out in March and April 1995 is described below. The initiator of the test was the Deutsche Gesellschaft für Klinische Chemie e.V., which is based in Bonn.

The material for testing consisted of two lyophilized samples of 25 ml urine each. The groups of substances to be analyzed are listed in Table 13-4.

Table 13-4. Substances and substance groups used in interlaboratory tests

Individual substances (I) and substance groups (G, I)	
Amphetamine and related substances (G)	Methaqualone/metabolites (G, I)
Amphetamine (I)	Opiates (G)
Methamphetamine (I)	Codeine (I)
Barbiturates (G, I)	Dihydrocodeine (I)
Benzodiazepines (G, I)	Morphine (I)
Cannabinoids (G, I)	Phencyclidine/metabolites (G, I)
Methadone/metabolites (G, I)	Propoxyphene/metabolites (G, I)
Cocaine/metabolites (G, I)	Tricyclic antidepressants (G, I)

In this interlaboratory test, the analysis was carried out using all appropriate techniques. Figure 13-7 shows the chromatogram of Sample A in non-acetylated (I) and acetylated (II) form. The amphetamine could only be detected after derivatization with acetic anhydride. In the acetylated sample, levallorphane and its metabolites were found. The presence of levallorphane can lead to a positive opiate test by the immunological method.

Figure 13-8 shows Sample A after acidic liquid sample preparation. Here, amobarbital could be determined.

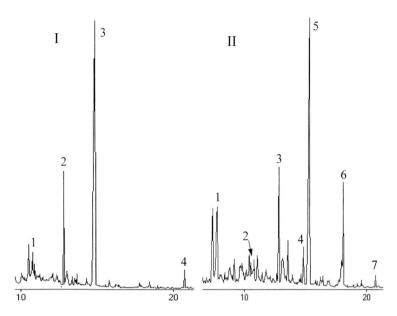

Fig. 13-7. Gas chromatogram of the interlaboratory test of Sample A. Alkaline sample
pretreatment by solid phase extraction.

     I – Sample A not acetylated; 1 = PCP; 2 = methadone; 3 = levallorphane;
       4 = alprazolam.

     II – Sample A acetylated; 1 = amphetamine AC; 2 = PCP; 3 = methadone;
       4 = levallorphane; 5 = levallorphane AC; 6 = levallorphane
       metabolite AC; 7 = alprazolam

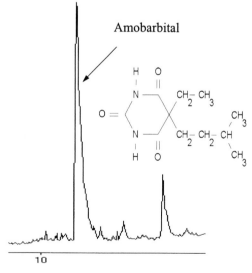

Fig. 13-8. Gas chromatogram of the interlaboratory test of Sample A. Acid sample
pretreatment with liquid-liquid extraction

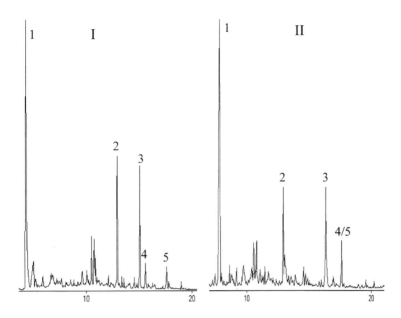

Fig. 13-9. Gas chromatogram of the interlaboratory test of Sample B. Alkaline sample
pretreatment with solid phase extraction.
I - Sample B not acetylated; 1 = phentermine; 2 = methaqualone;
3 = codeine; 4 = morphine; 5 = bromazepam.
II - Sample B acetylated; 1 = phentermine AC; 2 = methaqualone;
3 = codeine AC; 4 = diacetylmorphine; 5 = bromazepam

Figure 13-9 shows the chromatogram of Sample B non-acetylated (I) and
acetylated (II). Codeine and morphine were detected in the non-acetylated
sample. In the acetylated sample, the codeine was found in its acetylated form.
The morphine was converted into heroin.

THC was also tested for in all the samples by an immunological method.
However, the result was negative. Cocaine was identified in the SIM mode by
its characteristic masses 82, 182, and 303.

A list of institutions which organize interlaboratory tests is given in Table 13-5.

Table 13-5. Organizers of interlaboratory tests

Organization	Address
Deutsche Gesellschaft für Klinische Chemie e.V.	Postfach 150139 53040 Bonn
Gesellschaft für Qualitätskontrolle klinischer Arzneimittelanalysen und Toxikologie (KKGT)	Ido C. Dijkhuis Zentrale Krankenhausapotheke Postbus 43100 NL-2504 AC Den Haag
Cardiff Bioanalytical Services Ltd.	Cardiff Medicentre Heath Park Cardiff CF-4 4UJ, UK
GTFCh Gesellschaft für Toxikologische und Forensische Chemie	Langrabenstrasse 74 61118 Bad Vilbel

References for Part IV

[1] Wichmann, H.E., Schlipköter, H.W., Fülgraff, G.M., *Handbuch der Umwelt-medizin* Landsberg/Lech: Ecomed, 1992.

[2] Thomas, L., *Labor und Diagnose*, Marburg: Med. Verl. Ges. 1992.

[3] Rick, W., *Klinische Chemie und Mikroskopie* 6th edn., Berlin: Springer-Verlag, 1990.

[4] Allgemeine Geschäftsbedingungen der Deutschen Bundespost Postdienst für den Briefdienst Inland - AGB BfD Inl.

[5] Data sheet from BIO-RAD on Urin Toxicology Screen Control Lyphocheck®, Bio-Rad Laboratories GmbH, München, 10/92.

[6] Data sheet from BIO-RAD on Urin Toxicology Screen Confirm Control Lyphocheck®, Bio-Rad Laboratories GmbH, München, 10/93.

[7] Data sheet from DRG Instruments, Kontrollurin No. 19, DRG Instruments GmbH, Marburg, 10/94.

Index